629.13255 LAN

ONE WEEK LOAN

PENGUIN BOOKS

FLY BY WIRE

William Langewiesche is an author and journalist. He is currently
Vanity Fair's international correspondent, having made his name
writing for *Atlantic Monthly*. His strong, evocative prose is used
to devastating effect on a range of issues. Before embarking on
a writing career he worked as a pilot for fifteen years from the
age of eighteen. He has been termed one of the leading writers of
The New New Journalism, a group of writers who have secured
a place at the centre of contemporary American literature, as Tom
Wolfe and The New Journalism did in the sixties.

ALSO BY WILLIAM LANGEWIESCHE

Cutting for Sign

Sahara Unveiled

Aloft

American Ground

The Outlaw Sea

The Atomic Bazaar

FLY BY WIRE

The Geese, The Glide,
The 'Miracle' on the Hudson

WILLIAM LANGEWIESCHE

PENGUIN BOOKS

PENGUIN BOOKS

Published by the Penguin Group
Penguin Books Ltd, 80 Strand, London WC2R ORL, England
Penguin Group (USA) Inc., 375 Hudson Street, New York, New York 10014, USA
Penguin Group (Canada), 90 Eglinton Avenue East, Suite 700, Toronto, Ontario, Canada M4P 2Y3
(a division of Pearson Penguin Canada Inc.)
Penguin Ireland, 25 St Stephen's Green, Dublin 2, Ireland
(a division of Penguin Books Ltd)
Penguin Group (Australia), 250 Camberwell Road, Camberwell, Victoria 3124, Australia
(a division of Pearson Australia Group Pty Ltd)
Penguin Books India Pvt Ltd, 11 Community Centre,
Panchsheel Park, New Delhi – 110 017, India
Penguin Group (NZ), 67 Apollo Drive, Rosedale, North Shore 0632, New Zealand
(a division of Pearson New Zealand Ltd)
Penguin Books (South Africa) (Pty) Ltd, 24 Sturdee Avenue, Rosebank,
Johannesburg 2196, South Africa

Penguin Books Ltd, Registered Offices: 80 Strand, London WC2R ORL, England

www.penguin.com

First published in the United States by Farrar, Straus and Giroux 2009
Published in Penguin Books 2010
2

Copyright © William Langwiesche, 2009
All rights reserved

The moral right of the author has been asserted

Portions of this book were previously published in *Vanity Fair*

Printed in England by Clays Ltd, St Ives plc

978-0-141-04674-7

www.greenpenguin.co.uk

This is the second book I dedicate to Cullen Murphy,
my editor and friend.
To Cullen Murphy, again.

Contents

FLIGHT PATH
of
US AIRWAYS
FLIGHT 1549
•
January 15, 2009

280

NEWARK
INTERNATIONAL
AIRPORT

95

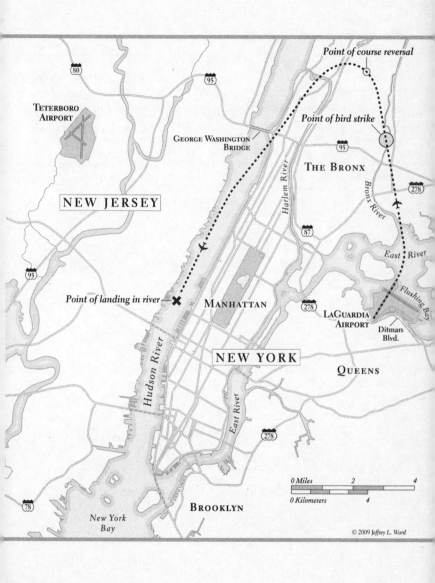

Point of course reversal

Point of bird strike

George Washington Bridge

THE BRONX

TETERBORO AIRPORT

Harlem River

NEW JERSEY

Bronx River

East River

Point of landing in river

MANHATTAN

Flushing Bay

LaGuardia Airport

Ditmars Blvd.

NEW YORK

QUEENS

Hudson River

East River

BROOKLYN

New York Bay

0 Miles 2 4

0 Kilometers 4

© 2009 Jeffrey L. Ward

FLY BY WIRE

New York, January 15, 2009

It was a wintry Thursday afternoon, and the city had turned inward on itself against the cold. On Manhattan's west side, a few people who happened to be looking toward the Hudson River caught a glimpse of an airline accident that initially brought back memories of another case, eight years earlier, of airplanes crashing into the heart of New York. This time it was US Airways Flight 1549, an Airbus A320 that ran into a flock of geese, lost thrust from both engines, and glided without power to a safe landing in the Hudson's frigid waters. The Department of Homeland Security flashed its badges, but only as bureaucracies do. There were no foreign terrorists here. The geese were innocent birds. The captain was the very definition of a good citizen, a man named Chesley Sullenberger whose life until now had been so uneventful that many of his peers at US Airways had overlooked his presence. Overnight he became a national hero as politicians, the press, and the public caught on to the man who would become known as "Sully."

It had snowed that morning, but the skies had largely cleared. The Airbus had departed from nearby LaGuardia Airport and had reached only 3,000 feet before being forced from the air.

From takeoff to splashdown, the flight had lasted just five minutes. Surveillance cameras on the shores captured the final moments from a distance. The airplane streaked onto the scene with its landing gear up, approached the river at a shallow angle, and settled into the water with a brief plume of spray. It swerved slightly to the left and came to a stop, floating nose high, drifting downriver at the speed of the current. Its tail soon sank below the surface, but the cockpit and much of the cabin remained dry. Within seconds the forward doors opened and two slide rafts inflated, one on each side. People began to pile into them and to emerge through the emergency hatches to stand on the wings. There were a hundred and fifty passengers and five crew members aboard. Relatively few wore life vests. None was dressed for the occasion. The airplane settled lower, until people standing on the wings were up to their waists in the cold water. But the first of several small ferries and rescue boats arrived four minutes later, and soon afterward everyone was safe.

Five minutes of flight. Four minutes until survival was assured. This book is the story of a short slice of life.

THE INQUEST

In June 2009, six months after Chesley Sullenberger struck a flock of Canada geese and glided his wounded US Airways Airbus to a successful ditching into the Hudson River, a public hearing on the case was held in Washington, D.C. It was organized by the crash investigators of the National Transportation Safety Board (NTSB), a small and independent federal agency that is renowned for its technical expertise. During the six months since the accident, the investigators had been dissecting the case and studying the factors behind it. Despite Sullenberger's skillful flying and the survival of everyone aboard, it turned out that there was much to consider here. Simply put, the successful outcome had been a very near thing. Furthermore, NTSB investigators are professional worriers. On the occasion of this hearing, they were going to release the information they had gleaned and, under the guise of taking sworn testimony from expert witnesses, publicize some of their concerns. What can be done about flocking birds, about jet engines, about water landings, about passenger briefings, about life rafts, about never again requiring people to stand on sinking wings to keep from drowning? What can be done about never again depending on such a chain of good luck?

The NTSB is meant to be pure, the speaker of truths no matter how impractical they may be. As an agency it is built that way. It cannot write regulations, mete out fines, impose technical standards on designs, or force its opinions on its fellow government bureaucracies. It does have the power of subpoena and can swear in people to encourage them to tell the truth, but this is more for show than for meaning. Rarely have people been prosecuted for lying to the NTSB, though people have lied to it plenty of times. In the end it really only has the power of persuasion at its disposal. Some on the staff call this the power of the raised eyebrow. Their highest hope is for incremental progress measured in years. That was to be the purpose of the hearing now. For two full days and part of a third, the NTSB was going to engage with a parade of pilots, officials, and engineers, few of them able to speak in clean English, and most of them wanting to make opening statements using PowerPoint displays. The standard stuff. The facts were known. For the audience it would be rough going, with no coffee allowed.

Those were sultry days in the capital. The sun cooked the haze. Every night thunderstorms roiled the skies overhead. There were probably twenty such hearings happening in the city. This one was held at the NTSB Conference Center, a windowless auditorium two levels below the street, across from the Smithsonian Institution, in the hotel and office complex called L'Enfant Plaza. You could certainly feel safe there. Finding it the first morning required navigating through an underground shopping arcade among subdued office workers streaming in from the connected Metro station, most wearing identity tags on nylon straps around their necks. You endured the crowd, descended a narrow escalator behind people who could not be bothered to walk down it, and finally came to the auditorium after passing through a security check manned by uniformed guards of Washington's skeptical underclass. Later, in a private moment, I asked several of them if they did not want to sit in on the proceedings, and they laughed. They said they preferred to stay in the anteroom and talk about television.

The auditorium had a sloping floor, and comfortable seats for 350 people. It was about half full for the first few hours, and nearly empty by day three. Presiding over the proceedings was an NTSB board member, a former US Airways pilot and safety expert named Robert Sumwalt, who, as it turned out, had once flown the very same airplane involved in the crash. Sumwalt is an avuncular Southerner with a vague or distracted manner, and he seemed to have trouble tracking some of the testimony that followed. His role was largely ceremonial anyway. He sat on a raised platform at the front of the room, flanked by two senior staff members, with assistants seated behind. Along the left wall, another raised platform accommodated two rows of "technical staff," the accident investigators who had done the work and who would conduct the principal questioning. Across from them, along the right wall, was another raised platform, where the witnesses would sit while testifying. Between these three platforms, at tables in a well, sat teams from the officially admitted parties—various players deemed to have a stake in the public record of the proceedings. They represented the Federal Aviation Administration (FAA), the manufacturers of the Airbus and the engines, the flight attendants' union, the US Airways pilots' union, and US Airways itself. It was understood that these people had agendas that were largely self-protective, but which they would express only implicitly, and in earnest terms of public safety. They, too, would have a chance to question the witnesses for the record.

The hearing started on time. Sumwalt read an opening statement, explaining the proceeding in general terms, dismissing any conflict of interest that as a former US Airways pilot he might appear to have, and finishing with a request that people take note of the exits from the room for use in the event of an emergency. Apparently he thought you just can't be too careful in life. That was the tone of the entire hearing.

The chief investigator led off with a bare-bones summary of the accident: it occurred on January 15, 2009, at 3:27 p.m.; there were 150 passengers and five crew members aboard; they were in an Airbus A320 bound from New York's LaGuardia Airport for Charlotte, North Carolina; the time from liftoff to the bird strike was 1 minute, 37 seconds; the birds were Canada geese at 2,700 feet; the geese caused a nearly complete loss of thrust by wrecking both engines; the glide to the river lasted 3 minutes, 31 seconds; the total flight time therefore was 5 minutes, 8 seconds; after the water landing the first rescue boat arrived in 3 minutes, 45 seconds; one flight attendant and four passengers were seriously injured; there were no fatalities.

Then the questioning began.

Sullenberger was the first up—at age fifty-eight, a tall, trim, white-haired man with a clipped white mustache, who seemed a bit overdressed for the hearing, in an elegant dark suit with a handkerchief in the breast pocket. He had arrived at L'Enfant Plaza in a limousine, accompanied by handlers from the pilots' union, and had entered by a side door to avoid the press. Not that he was averse to publicity. During the period since the accident, he had engaged one of the top publicity firms in San Francisco, near his home in suburban Danville, California, and he had made many appearances—accepting awards left and right, attending Barack Obama's presidential inauguration, standing with the crew for an ovation at the Super Bowl, throwing out the opening pitches at baseball games, mixing with movie stars at a *Vanity Fair* party, and sitting for interviews on national television. He had also signed a $3 million deal with HarperCollins to write two books, the first to be an inspirational autobiography coauthored by a bestselling personal-advice columnist for *The Wall Street Journal*, and titled *Highest Duty*. About the book's content, the publisher had said, "Sully believes his life experience prior to the emergency landing was a preparation for the success. And that life's greatest challenges can be met if we are ready for them." The publisher's statement had evoked sardonic comment nationwide, as had an Internet rumor—false—that the second book would contain inspirational poetry. Except among his most devoted fans, hero fatigue was setting in. The comedian Bill Maher captured the mood on HBO by showing a picture of Sullenberger waving to a crowd, underscored by a caption reading "Pompous Pilot." Maher said, "New rule. One more victory lap, and then you really have to get back to the cockpit."

It was funny but unfair. People who thought that Sullenberger had lost his bearings were underestimating the man. In private he remained the same person he had been before, not pompous at all, and so quiet about himself that at times he could seem shy. Intellectually he was the equal of the observers who thought he was grandstanding, and he knew as well as they did where he stood on the American scale. He had been diligent as a boy, and had become a diligent pilot. The career had certainly narrowed his experience in life. But he nonetheless possessed an attribute that those who mocked him had overlooked: he was capable of intense mental focus and exceptional self-control. Normally these traits do not much matter for airline pilots, because teamwork and cockpit routines serve well enough. But they had emerged in full force during the glide to the Hudson, during which Sullenberger had ruthlessly shed distractions, including his own fear of

death. He had pared down his task to making the right decision
about where to land, and had followed through with a high-stakes
flying job. His performance was a work of extraordinary concentration, which the public misread as coolness under fire. Some
soldiers will recognize the distinction.

Sullenberger maintained his concentration through the water
landing, the evacuation of the airplane, and the brief boat ride to
shore. Then a strange thing happened to him. He was no Charles
Lindbergh seeking to make history, no Chuck Yeager breaking
the speed of sound. The Übermensch era of aviation had long
since faded. But he crashed during a slump in the American
mood, and overnight he was transformed into a national hero, at a
time when people were hungry for one.

At that point he began to concentrate again. After decades of
enduring the insults of an airline career—the bankruptcies, the
cutbacks, the union strife, a 40 percent reduction in salary, the
destruction of his retirement pension—he was determined to leverage this unexpected opportunity to maximum advantage. He
was due to retire in seven years, at age sixty-five. Now he was suddenly on a ride as critical to his family as the glide to the river had
been, but mirrored upward, and with a destination less easy to
discern. They would go where the ride took him, his athletic wife
and their two teenage daughters with college ahead. Sullenberger
said he was moved by the flood of mail he had received, and was
glad to serve as an inspiration to so many people. Probably that's
right. But he was not self-delusional—for instance, he ignored
some clamoring that he run for public office—and he seemed to
be focusing on two rather more practical goals. The first was financial stability. He was forthright about it from the start, when
he announced through the press that he would consider all offers
and possibilities. He was going to gain from this event, and why

not? The second goal was slightly less obvious. It was to promote a union argument, couched as usual in the language of safety, that holds that if pilots are not better paid, airline travel may become increasingly unsafe.

Sullenberger is a dedicated union man, as any self-respecting pilot at US Airways should be. In the month following the accident, he appeared before Congress with his entire crew, and after receiving a standing ovation from the staff and committee members, he shifted the subject. He said, "I am worried that the airline piloting profession will not be able to continue to attract the best and the brightest." His message was that successive generations of pilots willing to work for lower wages might perform less well in flight, and especially during emergencies. Sullenberger seems to believe this, but it is a questionable assertion, since it links financial incentive to individual competence, and ignores the fact that, with exceptions, the "best and the brightest" have never chosen to become airline pilots, at whatever salary, because of the terrible this-is-my-life monotony of the job. Furthermore, although unusual stupidity is often fatal in flying, the correlation between superior intelligence and safety is unproven, given the other factors that intrude—especially arrogance, boredom, and passive rebellions of various kinds. If you had to pick the most desirable trait for airline pilots, it would probably be placidity. But safety aside, no pilot of whatever mental capacity enters the profession expecting to see his income cut, particularly when airline executives continue to increase their own compensation, as they have. This is what Sullenberger was legitimately complaining about to Congress. Ever since airline deregulation in the United States in 1978, which did away with route monopolies and noncompetitive pricing, and especially since the terrorist attacks of September 2001, which had all sorts of profound effects on the industry, most major American airlines have been miserable places to work.

Two days after the Hudson River landing, accident investigators interviewed the pilots in New York's Marriott Downtown hotel, near the site of the former World Trade Center. They started with Sullenberger's copilot, Jeffrey Skiles, age forty-nine, a man with twenty thousand hours of flight time, who had briefly served as the captain of a smallish Fokker twin-jet but had essentially been relegated to the position of copilot for his entire career because of successive reductions in the company ranks. Like Sullenberger, Skiles has a mustache. He has an alert, pleasant face and an unassuming manner. He mentioned wryly to the investigators that he had become perhaps the most experienced copilot in airline history. He was new, however, to the Airbus, and had just gone through transition training after years of wearing grooves into the right seats of Boeing 737s. At the time of the accident he had only about thirty-five hours of Airbus time, and was flying his first four-day trip in the A320 and its kind. He had never flown with Sullenberger before and did not know him socially. He had been impressed by Sullenberger's competence in the cockpit, as Sullenberger had been impressed by his. But now, because of all the publicity, he was going to be forever linked to this story, and known as the second fiddle for the rest of his life.

An NTSB investigator summarized Skiles's career path. He wrote, "He learned to fly at age fifteen or sixteen. He flew for a cargo company, then a commuter airline, and then was hired by US Airways in April 1986. He was a [Boeing] B-727 flight engineer when hired, upgraded to copilot until the airline parked the B-727. He then went to the DC-9 until the airline parked it, the Fokker until the airline parked it, then the B-737. His decision to move to the Airbus was voluntary, his only voluntary move, although it came just ahead of the company parking the B-737s."

Skiles was straightforward about the experience. He told the investigators that he had never enjoyed even the most enjoyable part of the career, which most pilots agree is the training. He said that he had once gotten a letter of praise from the company's director of operations—for making "great" public address announcements to the passengers. That was about it. He had an unblemished record, as most pilots do. No, he did not fly except on the job. Private flying is very expensive. Over the past eight years he had suffered a 50 percent reduction in salary, forcing him to supplement his income by working as a general contractor on his days off back home in Wisconsin. As might be expected, he was angry about this. When the investigators asked him if he enjoyed working with US Airways, he answered flatly, "No one likes working with the company." He remained on the job because the alternatives were even worse.

I don't mean to imply that Skiles is a bitter man. He is not. But in this interview he was completely frank. When asked about teamwork in the cockpit during the glide, he said there was little need for it, and little was involved: he had started into the checklist to restart the engines, and Sullenberger had done the flying. The division was plain and simple, and pretty obvious at the moment. You could arrive afterward and call it an exercise in Crew Resource Management—sorry, I mean CRM—if you insisted on fixing things up with formal language. CRM is indeed a useful term. Until recently it stood for Cockpit Resource Management and pertained only to pilots, until someone realized that the *C* could stand for *Crew*, allowing flight attendants into the program. Entire industries are built on this sort of progress. But frankly the glide had been very short, with no space for elaboration. I mean, actually, fuck it, the pilots had simply flown the airplane—and what else were they going to do? Skiles did not quite say this, but it is what he meant. As for the training they had received in dual-engine failures, and the procedures for water landings, he said it was premised on the latest philosophy about taking time to assess emergencies, and so it had not helped at all.

Sullenberger's interview was subtly different. As Skiles had done, he spoke mostly about technical details: what he had believed about the aircraft systems, what his logic had been during the glide, which switches he had thrown and why. He was in a strong position to answer, having exhibited profound piloting skill. He freely admitted to his uncertainties, without the slightest sign of defensiveness. Nonetheless, he was clearly more aware of the political context than Skiles had been, and of course of the opportunities now suddenly arising. In retrospect, he was concentrating hard. When asked if the US Airways training had helped him to handle the emergency, he said absolutely it had, and he cited the principles of maintaining control, managing the situation, and (oddly, in this context) landing as soon as possible. He also credited the training in Crew Resource Management, the clear definition of duties, and the clear communication of plans. An investigator asked him how he liked working at US Airways. He answered that it is "a good company." The investigator asked if the company exerted "external pressure" on the crews. The question, though poorly phrased, was an invitation to expound on the corporate culture of the airline. Sullenberger certainly understood this. In his water-logged bag in the Airbus he had a library book, *Just Culture: Balancing Safety and Accountability*, about precisely such issues in the airlines and similar organizations. Nonetheless, he answered, "I'm not sure," and left it at that. He was watching his words. He was determined to make no mistakes. He went out of his way to praise Skiles and the flight attendants. At the end he made a statement for the record. He said, "I could not be more happy and pleased and gratified that we got 155 people off the airplane. And it was due to the professionalism of my crew: Jeff, Donna, Sheila, and Doreen." He did not want to speculate on how training might be improved. He was tightly self-controlled.

.

Six months later he responded in the same manner during the NTSB hearing in Washington, D.C., though with more forethought apparent in his words. He sat very straight and gave short answers without elaboration or drama. The main questioner was a youngish NTSB investigator named Katherine Wilson, who had a fresh Ph.D. in applied psychology from the University of Central Florida, with a specialization in Crew Resource Management. Sullenberger referred to her as "Dr. Wilson." She referred to him as "Captain Sullenberger." She asked him a few technical questions, but for the most part just threw him flowers.

"How do you think that your experience with over twenty thousand hours as a pilot helped you during this experience?"

"It allowed me to focus clearly on the highest priorities at every stage of the flight, without having to constantly refer to the written guidance."

"Looking back at the accident event, is there anything you would do differently, if you were faced with that situation again?"

"I think what we did, the situation we faced in the time we had, First Officer Jeff Skiles, and Flight Attendants Donna Dent, Sheila Dail, and Doreen Welsh, we did the very best we could. And I'm proud to have been a member of a highly experienced, highly trained team."

"What lessons do you think we can learn from this accident?"

"I think it's the importance of CRM, the importance of a dedicated, well-experienced, highly trained crew that can overcome substantial odds. And working together as a team can bring about a good outcome."

"Is there anything else you would like to discuss today that we have not asked you so far?"

"Just to reiterate my gratitude for such a good outcome on January 15, and the amazingly quick response of the first responders from New York and New Jersey."

"Great. Thank you."

A French investigator who had been seconded to the panel tried to get a bit more technical about the Airbus itself, a radical semi-robotic European design that was known to the investigators to have participated actively in the survival of the passengers. In private, some of the test pilots and engineers from the Airbus company had been seething for months over Sullenberger's silence on the subject. His refusal to mention the unique qualities of the airplane was understood as a partisan stand in the context of a long and painful history, in which the A320, the world's first semi-robotic airliner, had been vehemently opposed by the unions, because it is designed around the idea that computers fly better than any human can—and indeed, in some emergencies, should override the pilots entirely, and firmly assume command. This is a complex and emotional subject, since it goes to the heart of a profession already in decline. No one had dared to bring it up directly—or to call attention to the airplane's contributions—lest this be seen as an attack on Sullenberger and an attempt to diminish his accomplishment. In that, there would be no advantage to anyone. Not to the NTSB, US Airways, or Airbus itself—and certainly not to the union. Nonetheless, for many in that hearing room, it was a subject very much in mind.

The Frenchman seemed to be thinking about it, because he asked a question that remotely pertained to the airplane's design. He said, "Could you please explain to us how you did choose the airspeed when you tried to do this emergency landing?" It was pretty light stuff.

Sullenberger answered with jargon. He said, "Yes. As we were not configured for landing, we didn't have a reference speed displayed on the PFD that we could fly. So I chose to use a margin above VLS."

"Configured for landing" means full flaps and landing gear down. "PFD" stands for "primary flight display." "VLS" stands for "velocity lowest selectable."

There was a moment of silence.

The Frenchman probed no further. He said, "Thank you, Captain."

Dr. Wilson echoed his gratitude. She said, "Thank you, Captain Sullenberger." To Robert Sumwalt, she said, "Mr. Chairman, we have no more questions at this time."

It was the turn now for the officially designated parties to ask their own questions. A woman from the flight attendants' union led off. She suggested to Sullenberger that rather than announcing, "Brace for impact," as he had over the cabin address system, he should have announced, "Brace for water impact."

Sullenberger easily batted this aside.

She then led him through a series of questions pertaining to the fact that only two of the four life rafts in the airplane had been usable, and that even if they had been loaded to their maximum capacity, forty-five people would have been unaccommodated. She asked Sullenberger, "Where do you think the additional forty-five people would have ended up?"

Sullenberger showed no sign of annoyance. He said, "I think

that they would have ended up where they ended up. Or they would have had to remain inside the forward fuselage while awaiting rescue."

She said, "Okay, taking the scenario a little bit further, assuming that rescue had not arrived prior to the aircraft submerging, where do you think these additional forty-five people would have ended up?"

Sullenberger balked. He said, "I would hesitate to speculate any further."

So she speculated for him. She said that without rescue boats on the scene, after the airplane had sunk, those people might have ended up in the water. She emphasized the word *might*, and repeated it, as if she had carefully considered some alternative. Then she asked, "How long do you think, taking into consideration how cold it was out there, that passengers not accommodated in rafts would have been able to survive, in cold water . . . if rescue boats had not been very close?"

Sumwalt finally intervened. To Sullenberger he said, "Um, are you an expert in survivability in water?"

Sullenberger said, "Member Sumwalt, the answer is no."

Sumwalt said, "Okay, we'll defer that question."

The woman said, "Okay."

This passed for high drama at the hearing. The audience remained admirably calm. The Federal Aviation Administration went next. The questioner was a large man who proved to be one of the shrewdest participants in the process, but whose main purpose seemed to be to build an obscure defense against any esoteric implication that his agency might somehow have done something wrong. He asked a question about command authority: "How did the US Airways 'Captain's Authority' portion of the Flight Operations Manual play into the actions on this flight?"

Sullenberger answered as if he had been writing a book. He said, "The captain's authority, or autonomy, the ability to make independent judgments within the framework of professional standards, is critical to aviation safety. It is codified in our Flight Operations Manual that the captain is ultimately responsible, and the final authority to all matters of flight. The buck stops here. And so we have the independent ability to make the right choice, to do the right thing every time, despite the occasional production pressures."

Airbus was next up. The team was fronted by a man of obvious intelligence who seemed like a slick Washington lawyer, but turned out to be something of a star pilot himself. He was American. Others on the team were French, American, and German, and included engineering test pilots who were intimately familiar with the airplane and its systems. Their expressions were guarded. For months they had pored over a wealth of information extracted from the airplane's flight data recorder, and they had run multiple simulations of the glide. They knew that the airplane's flight-control computers had performed remarkably well, seamlessly integrating themselves into Sullenberger's solutions and intervening assertively at the very end to guarantee a survivable touchdown. The test pilots believed that the airplane's functioning was a vindication of its visionary design. But they were not going to bring it up. They were going to get through this hearing and be done. Their front man said, "Good morning, Captain Sullenberger, but all of our questions have been answered by Captain Sullenberger, the technical panel, and the other party members. Thank you, sir."

Sullenberger said, "Thank you."

The engine manufacturer had no questions.

US Airways had no questions.

The pilots' union representative wanted to get back to crew resource management. There wasn't much to say. In fact, if you wanted to pick one accident in which elaborations on teamwork don't need to be made, this would be a good one to choose. It was I'll fly the airplane, you try to restart the engines. But crew resource management has become a central dogma, the sine qua non of airline flying, and because Sullenberger's landing had been successful, it seemed necessary to mix it in now. Sullenberger was willing to try. The union man asked him to describe his use of CRM that day, and Sullenberger said, "We had a crew briefing at the beginning of the trip, on Monday, January 12, where we aligned our goals, we talked about a few specifics, set the tone, and opened our channels of communication. So we functioned very well the entire time."

It was a valiant attempt. The union man seemed satisfied. The questioning shifted to the center stage, to each of the senior staffers on Sumwalt's right and left, and then to Sumwalt himself. Sumwalt was deferential. He said, "Tell me, in your mind, what made the critical difference in this event. How did this event turn out so well compared to other events that we see at the Safety Board?"

"I don't think it was one thing. I think it was many things that in aggregate added up. Again, we had a highly experienced, well-trained crew. First Officer Jeff Skiles and I worked well as a team."

It was time to let Sullenberger get on with his day, but Sumwalt was luxuriating in the exchange. He thanked Sullenberger for his analysis, and after a rambling preamble about some other case, he asked him what he thought about when driving to work before a trip. The answer was evidently not supposed to be unrelated to flying—his wife's exercise program, the need to pick up razor blades, annoyance with the offerings on TV.

Sullenberger said, "I think that one of the many challenges of our profession is that it has become so ultrasafe, where it's possible to go several calendar years without a single fatality, as we've just done recently, that it's sometimes easy to forget what's really at stake. Sometimes . . . we make it look too easy . . . So one of the challenges, I think, is to remain alert and vigilant and prepared, never knowing when or even if one might face some ultimate challenge."

This is what he thinks about when going to work? How often?

It was unfair to pose such questions to Sullenberger on the stand. But Sumwalt kept at it. He asked, "What can we extract from your mind-set, from the things you've learned, to hand over to others in the profession?"

You could almost hear the groans.

Sullenberger said, "I think it's important as we transition from one generation of pilots to the other that we pass on some of the institutional knowledge. No matter how much technology is available, an airplane is still ultimately an airplane. The physics are the same. And basic skills may ultimately be required when either the automation fails or it's no longer appropriate to use it."

At the Airbus table, people were listening with wooden faces, some staring down at their hands. They had had the grace to keep their peace before. But what design did Sullenberger think he had been flying? Nothing against him, but the automation in the accident airplane had emphatically not failed, and indeed had been integral to Sullenberger's control all the way down. Either he was mouthing generalities or this was a coded and familiar slap.

Anyway, he did not continue with it. He said, "In addition to learning fundamental skills well, we need to learn the important lessons that have been paid for at such great cost over generations. We need to know about the seminal accidents, and what came out of each of them. In other words, we need to know not only what to do but why we do it. So that in the case when there's no time to consult every written guidance, we can set clear priorities, and follow through with them, and execute them well." He was patching his response together. Mentally he reached for the library book that had drowned in the airplane: *Just Culture: Balancing Safety and Accountability*. New York mayor Michael Bloomberg had given him another copy, along with a key to the city. Sullenberger seemed to have finished the book by now. He said, "I think also it's important to note that nothing happens in isolation, that culture is important in every organization. And there must exist a culture from the very top of the organization, permeating throughout, [one] that values safety in a way that it's congruent, that our words and our actions match. And that people feel free to report safety deficiencies without fear of sanction. So all these things must happen together. We must balance accountability with safety."

"Thank you. In your mind does US Airways have that culture of safety?"

"I think that they do, and we're working very hard to make it what it needs to be every day."

"Thank you. I want to follow up on that by asking, in an interview that you had with the Safety Board, the question was, are there any external pressures from the company, and you said, 'I'm not sure.' What did you mean by not being sure?"

Sullenberger trod carefully here. He said, "I think there are a few situations that can occur where a captain is questioned. And again we must balance accountability with safety. The captain's authority is a precious commodity that cannot be denigrated. It's the ability to do the job. It's the ability to maintain professional standards at the highest level. No matter how inconvenient it may be. So we have to work every day to make sure that's the case on every flight."

Sumwalt said, "I want to bottle your mind-set, and make sure that everybody is drinking from that same bottle."

There was more to come in later sessions, when witnesses started talking about Threat and Error Management (TEM), Advanced Qualification Programs (AQPs), and Task Saturated Cognitive Skills (for which there appears to be no acronym yet). On the second day, two Asian men in identical gray suits fell asleep side by side with their heads back and their mouths hanging open. During Sullenberger's testimony, at least, people were in the presence of a celebrity. In the end, Sumwalt asked Sullenberger for final thoughts, and he summoned the discipline to answer one last time. He said, "I think it is that paying attention matters. That having awareness constantly matters. Continuing to build that mental model to build a team matters."

"Thank you. Captain Sullenberger, I have no further questions. I want to thank you very much for your testimony, for being here this morning, and for representing the piloting profession as you do. You are excused from the witness stand."

Sullenberger had gotten through. He is a brave and decent man. He said, "Thank you, Member Sumwalt," and soon made his escape.

Part One

GEAR UP

ONE

THE TAKEOFF

January 15, 3:25 p.m.

Chesley Sullenberger and Jeffrey Skiles met for the first time late in the afternoon of Monday, January 12, at the US Airways hub in Charlotte, North Carolina. Each had arrived from home, respectively in California and Wisconsin, by hitching rides on available flights. The two were paired for a four-day trip in various airplanes that they would swap with other crews as they proceeded, in order to keep the airplanes in nearly constant motion, for revenue generation and efficiency. Crews cannot be treated the same way. This was to be Sullenberger's first run in nearly two weeks, and he was well rested. Over the previous year, he had logged approximately 770 hours of flight time, an average of 16 hours a week, not counting the additional duty time on the ground, or the frequent transcontinental commutes—necessary because he chose to live so far from his assigned base. Skiles had flown at about the same leisurely pace, though also commuting long-distance to work. He, too, was well rested. However poorly paid flying for the airlines has become, it allows for a lot of relaxation, or at least time spent at home. Indeed, it would be a particularly gentle profession, as it was before, were it not for the insecurity and turmoil that have followed the industry's deregulation.

Among airlines that have survived, the turmoil has been no-where worse than at US Airways. The company went into Chap-ter 11 bankruptcy in August 2002, and was able to hang on only because of government loan guarantees—part of the huge pack-age of bailouts awarded to the airlines in the wake of the terrorist attacks of 2001, when air travel declined and financial mayhem ensued. US Airways then embarked on a campaign to slash costs by reducing its fleet, furloughing pilots, cutting salaries, eliminat-ing pensions, and doing away with free meals on its flights. The mood was reflected at the time in a somehow desperate slogan: "Get On Board." Please, goddamnit. The airline emerged from bankruptcy in 2003, only to be forced back into Chapter 11 a year later, in September 2004, as a result of high fuel prices and dead-locked negotiations with the pilots' union. Afterward, employees had to make concessions again, and they were bitter about it.

US Airways cut its labor costs by $1 billion following the sec-ond bankruptcy, and brought its overhead closer to that of pared-down airlines like Southwest. Sullenberger later referred to the effect in his congressional testimony, when he spoke out against "airline management teams who have used airline employees as an ATM." The airline executives Sullenberger had in mind were surely his own. In fairness, their hands had been tied by bankers imposing conditions for loans. The bankers in turn were eyeing the realities of a competitive market that is extremely sensitive to pricing, and in which customers, informed by the Internet, ag-gressively seek the best deals around. Morale at US Airways sank so low that during a Christmas snowstorm in 2004, angry flight attendants and ground personnel called in sick, causing the can-cellation of several hundred flights, snarling traffic nationwide, and resulting in the stranding of many thousands of passengers. The airline blamed the weather. The government blamed mismanagement. US Airways seemed truly to be dying. The airline business in the United States does not exist on the rational calculation of gain so much as on inertia and fascination. For whatever reason, US Airways once again was saved. It happened in 2005, in the nick of time, when the Phoenix-based America West Airlines took over, assuming US Airways's name, assets, and debt and allowing it to emerge once again from bankruptcy. America West had its own history of troubles, having gone through bankruptcy in the 1990s, and requiring a government loan of $380 million in the aftermath of the 9/11 attacks. This was an airline so close to the brink that it resorted to selling tray table advertising on its flights, and thought of this as an important innovation. It was nonetheless considered to be well run, and was able to bring big investors to the deal, including Airbus in Europe. The US Airways management team was fired, the America West name disappeared, and the now-former America West managers took over, moving the US Airways headquarters to their longtime Arizona base.

On January 12, 2009, when Sullenberger and Skiles met in Charlotte for their assigned four-day trip, the last of America West had recently disappeared from public view beneath the US Airways veneer. Whether for nostalgia or as a reminder of what had really happened and who was really in charge, the new radio call sign for all US Airways flights was the old "Cactus" of America West. Sullenberger did not approve, and he came up with a reason why. He said that during operations overseas—in Asia, Europe, and Latin America—foreign crews hearing "Cactus" on air traffic control frequencies might not correlate it to the US Airways paint scheme on the airplanes in sight, and so safety might be compromised. More likely, he simply resented the name.

Behind the façade of a unified airline, a war had broken out between the 3,400 original US Airways pilots and the 1,800 pilots of the former America West. These groups were known respectively as East and West. Their fight was about how to integrate the ranks and endorse a single unified contract with the company. Pilots in the East group (such as Sullenberger and Skiles) insisted that seniority be based purely on the date of hire, while pilots in the West group (who typically were newer to the profession) insisted that they had not bailed out US Airways only to drop to the bottom of the scale. It was a significant fight, because the ranking of pilots governs the terms of their jobs, including pay, schedule, routes, and the airplanes they fly. In the fracas, the East pilots had forced the entire lot to pull out of the once-powerful national union, which seemed to have sided with the America West crowd, and had formed a company-specific bargaining unit they called the US Airline Pilots Association. (This is the union that handled Sullenberger after the crash and was represented as an official party at the NTSB hearings in Washington.) The West pilots had reacted by forming another group, appropriately named the America West Airlines Pilots Protective Alliance. For three years now, these two groups were going at each other in court, working under separate contracts, and refusing to integrate in the cockpits. It was a shame, and they were all weaker for it. They were working in a bare-bones industry, and fighting over scraps.

You dealt with it as you could. You got by in life. Skiles had gone into the house-building business presumably because he had some knowledge in that area. On the website for the company he formed, he wrote, "Skiles Builders LLC is committed to building affordable, elegant homes. Our personal involvement and pride of workmanship ensure a superior product. Our homes are designed and constructed with both classic design and practical usability in mind. The highest quality products, skilled craftsmanship and exceptional detailing produce a home with a character and personality uniquely your own. From vision to reality we make your dreams happen."

As for Sullenberger, he had hung out a shingle as a safety consultant and had founded his own company, Safety Reliability Methods, Inc., behind a website in which the "About Us" section makes it clear that the "us" is him alone. At the start of a two-page résumé, he describes himself as follows:

EXECUTIVE SAFETY/RELIABILITY
MANAGEMENT PROFESSIONAL

BOTTOM-LINE DRIVEN MANAGER SUPPORTED BY PRO-GRESSIVELY RESPONSIBLE EXPERIENCE ACROSS 40+ YEARS IN THE AVIATION INDUSTRY. POSSESS IN-DEPTH UNDERSTANDING OF AVIATION OPERATIONS ACQUIRED THROUGH REAL-WORLD FLIGHT EXPERI-ENCE, PROFESSIONAL TRAINING AND LEADERSHIP ROLES WITH ONE OF THE WORLD'S LEADING AIR-LINES. HISTORY OF ACHIEVEMENT IN SAFETY, INNO-VATION, CREW TRAINING, OPERATIONAL IMPROVEMENT, COST SAVINGS, PRODUCTIVITY IMPROVEMENT AND CUSTOMER SERVICE. COMBINE STRONG INDUSTRY KNOWLEDGE AND BUSINESS LEADERSHIP SKILLS TO CONSISTENTLY MANAGE COMPLEX SCHEDULING, LEAD HIGH-PERFORMANCE, MOTIVATED TEAMS AND IMPLEMENT EFFICIENT PROCESSES THAT ENSURE SMOOTH OPERATIONS AND QUALITY CUSTOMER SER-VICE. STRONG COMMUNICATOR, EFFECTIVE NEGO-TIATOR AND MOTIVATIONAL TEAM BUILDER; ABLE TO EFFECTIVELY COMMUNICATE NEEDS AND MERGE

DISPARATE TEAMS IN THE SUPPORT OF MARKET OB-
JECTIVES. RESPECTED FOR WIDE RANGE OF INDUS-
TRY KNOWLEDGE, SOLID SENSE OF INTEGRITY AND
DEMONSTRATED PASSION FOR INDUSTRY AS A WHOLE
AS EVIDENCED BY LIFELONG CAREER OF FLYING.

In other words, he was an airline pilot. His need to compensate for the loss of income was painfully evident in the enterprise. There was something endearing in the very rigidity of the language, and in a large photograph on the website that showed him smiling in his airline pilot uniform, with captain's stripes on the sleeves. He was obviously a decent man. He was straining to broaden out. He had landed an affiliation as a visiting scholar at the University of California, Berkeley, at the Center for Catastrophic Risk Management—a construct that seems to have been designed for the purpose of hunting grants. Maybe the affiliation would help.

Sullenberger and Skiles certainly had time for these secondary pursuits, however unexpected in their lives. But on the afternoon of January 12, when they joined up in Charlotte, North Carolina, they set their financial concerns aside to do their job. Both men were feeling cheerful. They met the three flight attendants, Donna Dent, Sheila Dail, and Doreen Welsh. Donna Dent was the lead. She was a short-haired woman, age fifty-one, who had joined the airline in 1982 and had been flying with it for twenty-six years. Even more experienced was Sheila Dail, who at the age of fifty-seven still retained the looks once required for the job. She had joined the airline in 1980, as Sullenberger had, more than twenty-eight years earlier. But the real veteran of the crew was Doreen Welsh, now fifty-eight, who had joined US Airways when it was called Allegheny and she was twenty years old, in the dim and distant past of 1970. For reference, in 1970 Richard Nixon was in his first term in office, the war in Vietnam was raging, U.S. forces invaded Cambodia, protesters were shot dead in Ohio at Kent State University, Jimi Hendrix died young, and Barack Obama was nine years old. Furthermore, airline deregulation was still eight years ahead. However understandable Sullenberger's laments may be about the loss of flight crews' income, it must be said that he—like Skiles, Dent, and Dail—joined the airlines after the deregulation of the industry, when it was obvious that the unions would eventually be undermined by market forces, and that the unnaturally high salaries at the time simply could not be sustained. In that sense, among the five members of the crew, only Doreen Welsh could make a legitimate claim to having been blindsided by history. Incredibly, she had hung on throughout the ordeal, and had walked the aisles for thirty-eight years. You get the picture. Between the pilots up front and the flight attendants in the cabin, this was not a crew you wanted to complain to about the peanuts.

On the first leg of the trip, they hauled a load of passengers from Charlotte to San Francisco, a six-hour flight, which put them on the ground in California at 9:19 p.m. local time. Sullenberger drove home to Danville and went to bed at 11:00. He says he is regular about sleep, and good at it. He likes eight hours to feel rested. In the morning he rose at 7:00 and had breakfast with the children. Four hours later he left the house and drove to the airport for a 12:20 show time. Skiles had spent the night in a hotel, and had walked for about an hour before going to sleep. In the morning he had gone for another walk, for five or six miles, before returning to the hotel room and catching a ride to the airport on time. Presumably the flight attendants had enjoyed equally restful stays.

It was January 13, the second day of the trip. They picked up their assigned Airbus at the gate, loaded the passengers, took off from San Francisco at 1:15, and crossed the country to Pittsburgh in just under five hours. They landed at 9:00 p.m. eastern time and went to a hotel by the airport for a short ten-hour layover. They woke up early, took a 6:00 a.m. van to the airport, flew a flight to New York's LaGuardia Airport, and, after a delay there, flew back to Pittsburgh with a typical New York load of provincial tourists and burned-out business travelers. That was their workday. It was January 14. Because they landed early and had a long layover until the next morning, they went to a hotel downtown for the distractions of the city. It was snowing. Sullenberger took a walk, ate dinner alone, answered emails, and went to bed early. Skiles went to see a movie—*Gran Torino*, Clint Eastwood; it was really good. He had nothing alcoholic to drink, and had not had for ten years. Sullenberger was more of a drinker: he had had a beer ten days before. Skiles returned to the hotel and slept.

The day of the accident, January 15, was the fourth and last day of the trip. The crew left the downtown hotel at 7:30 in the morning in a van. At the airport Sullenberger ate a banana and a raisin bagel with cream cheese. Skiles ate nothing, which was normal for him at that hour. Their first run was to Charlotte, in a new stretched Airbus A321, which they were eager to fly. After pushing back from the gate, they had the airplane deiced. They lifted off from Pittsburgh at 8:56 in the morning, and two hours later they landed, after a typically uneventful flight. In Charlotte they switched airplanes for a scheduled flight to LaGuardia. The assigned airplane was a 150-passenger Airbus A320, about nine years old, a veteran of 16,298 flights and 25,239 hours of operation. Two days earlier, on a flight from LaGuardia, its right engine had burped because of a faulty temperature probe. The airline's mechanics had replaced the probe. The airplane was in excellent shape.

Skiles had a slice of pizza in the Charlotte terminal before settling with Sullenberger into the cockpit. The departure was slightly delayed because of snowfall from a cold front passing over New York, but at noon they lifted off from Charlotte. Crews usually alternate duties during trips, with one pilot and then the other doing the principal flying, and normally it would have been Skiles's turn to fly. But because he was still new to the Airbus and not yet authorized to land on runways contaminated with slush or snow, Sullenberger took the run. It was something over two hours long. By the time they got to New York, the cold front had passed, the snow had stopped falling, and the skies were rapidly clearing. The temperature at LaGuardia was twenty-one degrees Fahrenheit, and a brisk north wind was blowing. The visibility was superb. The runways were dry. Sullenberger landed the airplane and taxied to the gate at 2:23 in the afternoon.

•

At LaGuardia, the flight attendants disembarked their passengers and began to take on more—a full load of 150 people, who had been milling around and waiting as usual. They formed an average crowd for the airline, including ninety-five men, fifty-two women, two little girls ages six and four, and a baby boy nine months old. The baby did not have a child seat, and would sit perched unrestrained on his mother's lap. The father was also on the flight, as was another of the family's children, the four-year-old girl just mentioned. The family could not sit together, and made a fuss about it. The father and daughter ended up several rows behind the mother and son. The cabin was full except for a middle seat in the last row. Twenty-three passengers were Bank of America employees. One was an Australian folk singer. Two were pilots for other companies, riding for free. The oldest passenger was a woman of eighty-five, who moved with difficulty and needed a walker. She was accompanied by her daughter, who was fifty-eight. The average age of the men was forty-two. The average age of the women was forty-four. The shortest passenger was the baby boy, who measured 2 feet 5 inches long. He was the lightest aboard, at 23 pounds. The tallest passenger was a man of 6 feet 6 inches, who weighed 230 pounds. Somehow he squeezed into a window seat in the economy section. The heaviest passenger was a woman of 5 feet 4 inches who weighed 293 pounds. She squeezed into an aisle seat farther back. The average size of the men was 5 feet 11 inches, at 191 pounds. The average size of the women was 5 feet 5 inches, at 148 pounds. These are close to the averages for adult Americans. The boarding was pretty standard, too, with people blocking the aisle and cramming things into the overhead bins, one seat double-assigned, and flight attendants with barely repressed impatience urging people to hurry and sit down. While these unpleasantries transpired, Skiles did a routine walk-around of the airplane outside in the cold, and as usual found nothing wrong. Sullenberger bought an average sandwich in the terminal and went back to the cockpit, expecting to eat in flight. This was to be the final run of the four-day trip, a return to Charlotte as Flight 1549. It was Skiles's turn to do the primary flying, though with Sullenberger still ultimately in charge.

They had battled 185-mile-per-hour headwinds coming north to New York, but would gain advantage now by riding the same winds south. Skiles expressed amazement that the difference in flight time would be nearly an hour. The comment was conversational. It appears near the start of the official transcript of the cockpit voice recording recovered after the accident. The time was 3:03 p.m., and they were ready to be pushed back from the gate. The cabin doors were closed, and the emergency slides were armed to inflate automatically if the doors were opened. One of the flight attendants addressed the cabin. "If everyone would please take their seats . . ." Shortly afterward a flight attendant came to the pilots in cockpit and said, "Seated and stowed."

Sullenberger said, "Thank you, all set." The flight attendant shut the cockpit door, but it did not latch. Sullenberger said, "Okay, that darned door again."

Actually, it's not clear that he said "darned." The transcript inserts a prudish hash mark, "#," in the place of expletives. I guess he might have called it a "damned" door, but given his cockpit demeanor, it is hard to escalate beyond that. A goddamned door? Unlikely. A fucking door? It's inconceivable, though *fucking this* and *fucking that* is common cockpit talk.

Skiles said, "What's wrong?"

"This."

"Oh."

"You have to slam it pretty hard."

Apparently the flight attendant slammed the door. Afterward, no terrorist could breeze in unannounced. Referring to the reported weather, Sullenberger said, "Got the newest Charlotte." The weather there was fine.

In the cabin a flight attendant said, "Ladies and gentlemen, all electronic devices have to be turned off at this time." They were still hooked up to the tug and rolling backward on the ramp. LaGuardia was using Runway 31 for arrivals, and Runway 4 for departures. Runway 31 points to the northwest, on a compass heading of 310 degrees. Runway 4 points to the northeast, on a compass heading of 40 degrees. The runways cross. When LaGuardia is operating according to this plan, each airplane takes off after another one lands. Landing airplanes have priority for obvious reasons, but it is important not to let too many of them in without releasing others to fly, because LaGuardia is a cramped airport, rapidly overloaded by accumulations on the ground. That is the controllers' concern more than it is the pilots', though both groups have efficiency in mind. Safety is an underlying issue, but with that already in hand, what counts is to keep the passengers moving. In and out, in and out, there are only three major airports for New York, and among them they have to keep the great city alive. The threshold to Runway 31 is close to the US Airways terminal, and the threshold to Runway 4 is distant. Sullenberger said, "I was hoping we could land on four and take off on thirty-one, but it didn't quite work out that way."

Skiles said, "Well, we can make an attempt to beat Northwest here anyways." He was referring to another airliner getting ready to go.

Sullenberger is not the sort to be rushed. He said, "What's that?"

Skiles said, "So we can make an attempt to beat Northwest, but he's already starting, isn't he."

"Yeah. And we have to pull up before we can even start on this."

In the cabin a flight attendant gave the ever-expanding safety talk: the seat belts, the signs and lighted pathways, the eight emergency exits, two in the front, two in the back, and four over the wings, the four inflatable slides, the oxygen masks that will drop if needed, the do-not-hide-in-the-bathrooms-and-try-to-smoke-after-disabling-the-smoke-detectors, the thank-you-for-flying-our-miserable-airline. The passengers suffered through it for several minutes with varying degrees of indifference. The flight attendant explained that the seat bottoms could be used as flotation devices. Paradoxically, as it turned out, one thing she neglected to say was that because this particular airplane happened to be approved for extended offshore use, its emergency slides were of a sort that could be detached from the fuselage as life rafts, and every seat was equipped with a life vest stored in a pocket underneath. The omission was company policy, and it was not strictly illegal, because the run to Charlotte would not take Flight 1549 offshore, as offshore is defined in the regulations. For an overland flight such as this, the federal government required a briefing on the seat-bottom flotation, but only recommended one on the existence and use of life vests. US Airways was in a stingy mood. Indeed, the airplane had previously been equipped with a video system that briefed the passengers on all the safety equipment aboard, but the system had been disconnected to shave a few pennies from the overall operating costs, and flight attendants had been forced back out into the aisles—day in, day out, for reduced wages—with instructions to do the least to get by. This helps to explain why so many passengers on Flight 1549 ended up without life vests, standing on the sinking wings. For the management at US Airways, it was lucky that none of them drowned.

The pilots started the engines. In the cockpit this entailed rotating a knob to "Ignition/Start," and flipping each engine's master switch from "Off" to "On." The procedure was miraculously easy compared to those of the old manual starts. The pilots dealt with the left engine first, then the right. They monitored the indications of engine speed, fuel flow, and temperature. It took a couple of minutes. Skiles spoke of Delta's recent acquisition of Northwest Airlines. He said, "Wonder how the Northwest and Delta pilots are gettin' on."

Sullenberger said, "I wonder about that, too . . . I have no idea . . . Yeah, hopefully better than we and West do."

"Be hard to do worse."

"Yeah. Well, I hadn't heard much about it lately. But I can't imagine it'd be any better."

LaGuardia Ground Control instructed them to taxi to Runway 4. While proceeding, Skiles confirmed the initial clearance after takeoff—an immediate left turn to a heading of north (360 degrees) and a climb to 5,000 feet. There were delays on the way to the runway. The pilots went through the brief pre-takeoff checks. Ground Control put them in line behind Northwest and told them to switch to the primary frequency, known as "Local," or simply "LaGuardia Tower." The Tower controls movements on the runways and in the immediate vicinity of the airport. Business that afternoon was fairly relaxed for LaGuardia, but would have been considered intense at most other airports. There are airports in some countries where the control tower is hardly more than an employment refuge: controllers in those places sit silent for hours (and sometimes days) before getting to say the obvious to some flight crew that happens along— "cleared for takeoff" or "cleared to land." At LaGuardia it's not like that. The controllers are virtuoso performers. When Sullenberger and Skiles switched over to the Tower frequency, the Local controller was not only handling inbounds and outbounds on crossing runways, but, for additional sport, was also interweaving the movements of three Port Authority snowplow teams. His name was Anthony Wajda. He had been a controller for eight years, and had transferred to LaGuardia less than three months before. He was not hesitant by nature. He considered traffic to be light. He was keeping New York alive. It sounded like this:

WAJDA: United 672, you can exit on Tango behind US Air, or go down to Sierra, your choice. Ground, point seven.

UNITED: Looks like we'll make Tango behind US Air. We'll call ground. United 672.

WAJDA: Thank you. American 753, cleared for takeoff Runway 4.

AMERICAN: Clear for takeoff, 753.

WISCONSIN (inbound): Tower, Wisconsin 3650 cleared to land . . .

WAJDA: Wisconsin 3650, 31, wind 010 at 10, traffic will depart off 4.

WISCONSIN: 31, cleared to land, Wisconsin 3650.

US AIRWAYS (inbound): 2131 over the tanks.

WAJDA: Cactus 2131, LaGuardia Tower, number two. I will have your landing clearance shortly.

US AIRWAYS: Roger.

WAJDA (to a snowplow): Team 3, you can proceed onto Runway 4. Just remain south of the intersection 31.

SNOWPLOW: Tower, uh, Team 3, uh, will like to go up, uh, Double Alpha onto the intersection, sir.

WAJDA: Ah, that's going to be a problem. We have too many arrivals right now, but that . . . You have some other thing you want to do first, until final lightens up?

SNOWPLOW: Uh, we'll just stand by, uh, or if you can give us, give us clearance onto, uh, 4, we'll do, uh, Papa.

WAJDA: Yeah, you can do Papa right now if you want to proceed onto 4 on Papa. Just remain south of 31.

SNOWPLOW: Roger.

SECOND SNOWPLOW: Tower, Team 2.

WAJDA: Team 2.

SECOND SNOWPLOW: Team 2 like to proceed on Runway 4
    at Fox.

WAJDA: Team 2, you can proceed onto 4 at Foxtrot.

SECOND SNOWPLOW: Team 2 proceeding.

WAJDA: American 753, contact Departure.

AMERICAN: See you.

WAJDA: Wisconsin 3650, Ground, point seven.

WISCONSIN: Good afternoon, 'Consin 3650.

WAJDA: Cactus 2131 cleared to land 31, wind 010 [at] 10.

US AIR: Cleared to land 31, Cactus 2131.

DELTA (inbound): Delta 1356 coming up the freeway.

WAJDA: Delta 1356, LaGuardia Tower, you can start reducing.
    You're about 50 knots faster than the Airbus ahead.

DELTA: Got the anchor out.

This was a mere two-minute interval in Wajda's day. He put
Northwest into position on Runway 4, ready to roll through the
first gap offered by the inbound traffic and the plows. By no means
was he yet working at his full capacity. One gets the feeling he was
simultaneously juggling eggs and maybe playing Scrabble, just to
limber up for the evening rush still to come. At 3:21 he put Sullenberger and Skiles into position at the top of Runway 4, where
he held them for four minutes while two airplanes landed on the
crossing runway and the snowplows continued to work. In the
cockpit Sullenberger said, "Your brakes, your aircraft." With this
he assigned the handling of the primary flight controls to Skiles.

Skiles formally agreed. He said, "My aircraft."

At 3:25 Wajda cleared them for takeoff.

They pushed the throttles forward.

Skiles said, "TOGA," for "takeoff go-around thrust."

Sullenberger said, "TOGA set." The airplane accelerated down the runway. Sullenberger called out a speed: "Eighty."

"Checked."

"V-one. Rotate."

Skiles eased his control stick back to raise the nose. The wings bit into the air, and the airplane lifted smoothly off the ground. Sullenberger confirmed the climb. He said, "Positive rate."

Skiles said, "Gear up, please."

Sullenberger retracted the landing gear. "Gear up."

Wajda radioed, "Cactus 1549, contact New York Departure, good day."

Sullenberger answered, "Good day." They were passing through 700 feet, accelerating through 230 miles per hour, and banking left to go north, on the initial climb to 5,000 feet. In the cabin some passengers had already fallen asleep.

TWO

THE BIRDS

3:26 p.m.

At about that time a flock of Canada geese was flying at 2,700 feet southwest-bound over the Bronx in a loose-echelon formation, tending to business as usual, with nothing special in mind. Much about those particular geese will never be known—for instance, where they had come from that day, and where they were headed, and why—but it is likely that they were well fed and self-satisfied. Evidently they were also fairly dumb. Their stupidity cannot be held against them, since they were just birds, after all, but geese are said to be adaptive creatures, and it is hard not to think that they should have had better sense than to go blithely wandering through New York City's skies. New York is a busy place, and January 15 was a typical day there, propelled by all those schedules to keep. Was that so difficult to understand?

By January 2009, a full century had passed since the first bird had been killed by humans suddenly flying around. Orville Wright was at the controls that first time, on September 7, 1908, and he chased the bird down over a cornfield near Dayton, Ohio. Since then, for birds, the situation had grown dire worldwide. From 1990 through 2007, in the United States alone, civil aircraft struck birds on several hundred thousand occasions, often killing multiples at a time. The toll leveled around 2002, apparently because of the decline in air traffic following the September 11 attacks, but this proved to be a temporary reprieve. By 2007 the slaughter had soared to record levels, and with it had come a tendency to blame the victims and persecute them on the ground. There are some six billion birds in the United States, every one of them an easy target. Persecuting them on the ground is known as "mitigation." It can take various forms, some no more than harassment, others, including lethal gassing, amounting to mass slaughter. Canada geese have become particularly vulnerable politically. Those above the Bronx that day were cruising just beneath the clouds. There were perhaps forty of them there, and possibly many more. They were expending minimal effort at staying aloft, finessing the vortices swirling from one another's wings and maneuvering to keep their companions in sight. In theory these are the purposes of goose formations. And nature is marvelous, of course. But about six miles north of LaGuardia, and at about the same time that Wajda switched US Airways Flight 1549 to New York Departure Control, the geese flew into the departure corridor from Runway 4.

Immediately after takeoff, Flight 1549 swept past the city's enormous prison complex on Rikers Island, off to the left. Climbing through 500 feet, Skiles rolled the Airbus into a 20-degree left bank, to begin the required turn from northeast to the north. Sullenberger checked in with Departure Control, saying, "Cactus 1549, 700, climbing [to] 5,000."

A controller answered, "Cactus 1549, New York Departure, radar contact. Climb and maintain 15,000." The voice was crisp. The clearance was routine. The controller worked in a windowless radar room fifteen miles to the east of LaGuardia, on Long Island, in a mixed industrial and commercial district of Garden

City. The facility there is known as New York Terminal Radar
Approach Control, or TRACON. Its function is to sort New York
arrivals into efficient streams lined up for the runways at the three
main airports (and a dozen secondary ones), and through those
arrivals to thread the full volume of departures. It is not a job for
the timid; and indeed the controllers there are considered to be
among the best in the world. The controller who answered Sullenberger now was certainly one of them, a marathon runner
named Patrick Harten, age thirty-four, who had studied biochem-
istry in college, but whose father had been a controller and had
persuaded him to give the profession a shot. Harten had quickly
discovered that he had an affinity for the job, and liked it best when
the pressure was on; he had been working at New York TRACON
since 1999. As necessary, Harten can be a quick talker. Once, when
he was working the Newark arrivals sector—probably the most
challenging air traffic control position in the United States—he
issued a rapid-fire clearance to a Southerner flying for Delta, and
the pilot drawled back, "Do y'all hear how slow I'm talkin'? That's
how slow I hear."

Controllers identify themselves for the record with two ini-
tials. Harten's are NY. He is married to a gentle woman named
Regina, who teaches remedial reading at a primary school on
Long Island. On January 15 they were going through the final
stages of a two-year process to adopt a child in Russia. They subsequently flew to Leningrad and returned with a blue-eyed baby,
Patrick Ivan Harten, who has become the center of their lives. On
January 15, that lay ahead. Harten went on duty about fifteen
minutes before Sullenberger's call. The airspace he was control-
ling had an irregular shape, starting at LaGuardia and stretching
in places to thirty miles. It extended from the surface to 12,000
feet above the airport and, starting five miles to the north, up to

15,000 feet, the highest altitude to which Harten could climb the airplanes under his control. When he went on duty, he set his radar display to a standard configuration, which filtered out most non-aircraft targets, including the geese above the Bronx. During the NTSB investigation it turned out that the geese did register with the facility's primary, or "raw," radar system, somewhere in the bowels of a record-keeping computer—but among a thicket of false radar returns and unessential clutter from which they were difficult to sort out. As is normal in air traffic control, Harten was not looking at that raw information, but rather at an enhanced and cleaned-up display that was dense enough already with airplanes. Birds? There are a lot of them in the New York City area, and they're always flying around. That was just assumed. Even if you could see them on radar, you would have no idea of their altitude. Over his ten years on the job, Harten had seen only a few on his displays, and on those exceptional occasions, he had known that they were birds only because of correlated reports from pilots in flight. In any case, on January 15 the geese did not show up on the screen, and no pilots reported them in sight.

Harten knew that Cactus 1549 was coming, because he got a call on the landline from LaGuardia Tower that the airplane was on the runway and rolling. When Sullenberger checked in, Harten cleared him for an uninterrupted climb to 15,000 feet, because there was no conflicting traffic ahead. Sullenberger acknowledged the new altitude and pulled the throttles back to the climb-thrust setting. To Skiles he said, "Fifteen."

Skiles said, "Fifteen. Climb." He rolled out of the left turn, heading due north. The airplane passed through 1,400 feet, accelerating through 185 miles per hour. The landing lights were on as a standard precaution to make the airplane more visible head-on.

Sullenberger said, "Climb's set." Forty-five seconds had passed since the airplane's liftoff.

Skiles called for a flap retraction from position two to position one.

Sullenberger said, "Flaps one." They flew in silence for twenty seconds. It was a beautiful day. The Hudson River stretched upstream, below and to the left, on the far side of the Bronx. Looking outside, Sullenberger said, "What a view of the Hudson today."

"Yeah." But Skiles was all business. He said, "Flaps up, please. [Do the] after-takeoff checklist."

Sullenberger said, "Flaps up." The checklist was short. He said, "After-takeoff checklist complete."

They had been airborne for a minute and a half. Skiles was flying the airplane manually, with the autopilot off. They were climbing through 2,650 feet at 250 miles per hour—and suddenly they came upon the geese. Both pilots saw them at nearly the same time—a flock in line formation, ahead, above, and slightly to the right. Skiles sensed that the birds would pass underneath the airplane. To Sullenberger's eyes, however, they seemed to fill the windscreen. There were many, and they were large. He later said that he should have ducked below the glare shield for safety, but there was no time. He said, "Birds!"

Skiles said, "Whoa!" and flinched.

•

One should never anthropomorphize geese, of course, but certain familiar realities for those birds can be surmised. The most obvious is that they were surprised. This raises the question of why. Geese have hearing similar to ours, meaning that beyond the air noise of their own flight, they may have been deaf to the oncoming Airbus, which, like other modern airliners, has hushed engines that buzz rather than roar, and are nearly silent when heard from in front. Even if the birds did hear the airplane, they may not have localized the threat. Localization is done with the eyes. As best as is known, geese have visual acuity about like ours; however, they possess much larger fields of view because of the placement of their eyes, which do not face forward, but sit on the sides of their narrow heads. Researchers who have looked into the matter (and there are very few) suggest that geese in flight may see their environment as a nearly complete globe, simultaneously from directly above to directly below, and through 330 degrees of the horizon, with overlapping binocular vision for about 20 degrees straight ahead and monocular vision for an additional 155 degrees to both the left and right sides. Furthermore, because geese lack foveae (the part of the eye in humans and birds of prey that is responsible for sharply focused central vision) it is believed that they may see everything with equal sharpness without having to move their eyes. This means they would see every word on this page simultaneously, though comprehension would be a problem.

Literacy, however, has never been a requirement for survival. The question remains, therefore, of why these geese allowed themselves to be killed. We know that the airplane did not sneak up on them from behind. They were flying southwest, and it was flying north, approaching from below, ahead, and to the left. Assuming that the birds cared about survival, this leaves three scenarios to ponder: The first is that they did not see the airplane because it was camouflaged against the city background. The second is that they saw the airplane, but because it was stationary in the field of view (as all objects on collision courses are) they did not recognize it as a threat, or even as a moving object, until sud-

denly it surged upon them. In this second scenario it is possible
that they may have been lulled by the somewhat limited depth
perception afforded by monocular (left-eye-only) views of the
oncoming airplane. The third scenario is openly anthropomor-
phic, but was suggested to me by an authority on birds in flight,
Dr. Frank Heppner, who teaches zoology at the University of
Rhode Island and is notable for many reasons, including his admi-
ration for Canada geese. In defense of their minds, he proposed
that they might have seen the Airbus coming, might have understood what it was, might have recognized its relative speed, and
for want of solutions, might have decided to hold steady and allow
the pilots to maneuver around them. Such tactics are rational under certain crossing conditions. Then Heppner went further. He
said that in their own manner the geese might also simply have
thought, "What the fuck! We have the right of way here!" He was
joking, sort of. He said if that is indeed what the geese were thinking, it means they were birds as dumb as certain sailors who insist
on tacking across the bows of oncoming ships on the open sea. But
he did not believe it about them—and he preferred not to dwell
on such failures.

Birds. There are some basic divides. Among those that engage in socially organized flight, the big species cruise in staggered formations, and the smaller ones behave quite differently,
maneuvering in synchronized swarms known to some ornitholo-
gists as cluster flocks. The term *cluster flock* is of course a nerdish
reference to *cluster fuck*, though cluster-fucking is chaotic by definition, and cluster-flocking is not. Also, cluster-flocking is a mys-
tery. It is not known why birds do it, or, more important, how they
coordinate their moves, but their behavior is currently of interest
to some mathematicians and computer scientists, as well as to certain nerdish war-gamers at the Pentagon. It ought to be of interest

to cabin safety specialists, too, because experience shows that in
the rush to escape from crashed airliners, some passengers cluster-
fuck when they should cluster-flock. But that's getting ahead of
the birds in the story. In the annals of encounters between airplanes and birds, it is difficult to discern which bird species are
safer—those that travel in formation, those that cluster, or those,
such as seagulls, that flock without obvious organization. What is
known is that all species of grouping birds will try to get out of the
way when they are confronted by airplanes in flight. Solo fliers
such as eagles and other raptors will usually do the same. Pilots
likewise will try to stay away from birds in flight. There are exceptions. At the redneck extreme are pilots who go gunning for eagles
from high-wing, open-sided airplanes like Piper Super Cubs. In
practice, it takes two rednecks to do the job—a pilot to fly tight
turns and a shotgunner who will not mistakenly blow off the airplane's wing strut or wheel. Raptors, for their part, can be aggressive, too. The annals document repeated occasions when they
have attacked small airplanes. Those raptors are not rednecks, of
course. They are more like suicide bombers. A woman who has
studied the phenomenon suggested recently to me that they may
suffer from excessive testosterone. She did not disapprove, as
other women do. Actually, she seemed to be impressed.

But Canada geese are not so admirable. Years ago they had a
good reputation in New York. They were visitors from the distant
north who graced the city each fall and spring, igniting people's
imaginations and providing essential connections to the vastness
of the continent beyond the confines of the streets. When they
passed overhead in their majestic formations they seemed bound
for faraway places. In the early 1960s, however, the situation began to change after state wildlife agencies came up with a bioen-
gineering scheme whose purpose in part was to enhance state

revenues by stimulating the purchase of bird-hunting licenses. The agencies captured breeding pairs of an endangered but supersize subspecies known as the Giant Canada goose and, by clipping their wings, forced them to settle permanently into authorized nesting grounds along the Eastern Seaboard and elsewhere in the United States. The offspring of the clipped-wing geese imprinted to the new locations and, having lost the collective memory of migration, became full-time resident populations—endowed, however, with the urge and ability to fly. Simultaneously, it seems, other Canada geese may have given up on migration simply in response to changes in farming techniques, which left a new abundance of corn on the ground in the Midwest and the Middle Atlantic states. Then came the effects of Rachel Carson's *Silent Spring*, the eventual banning of certain pesticides and chemicals harmful to birds, the imposition of environmental protection laws, and the associated gentrification of former farmlands in places such as Long Island, New Jersey, and Connecticut. The newly nonmigratory giant Canada geese settled comfortably into a paradise with few predators, where hunting was frowned upon, where food was abundant, and where there were plenty of golf courses, corporate lawns, and preserved wetlands to dominate. In the United States their population grew from about two hundred thousand in 1970, to one million in 1990, to four million today. Nationwide, they now outnumber their migratory cousins by a ratio of more than two to one. The residents are particularly magnificent in flight, partly because of their unusually large size and wings that span up to six feet wide. But they are also insatiable overgrazers and prodigious defecators, and they can be disconcertingly defensive about their young. In a shift of public emotion, they are no longer seen as honored visitors, but as vermin and pests. Those are dangerous categories for any living thing, and all the more so now for a population that to some extent can be called the product of misguided men.

An entire industry has grown up around people's annoyance with these creatures. The industry has been boosted by the delicious complication that all Canada geese, including the implanted resident ones, are protected by the Migratory Bird Protection Act of 1918, and the associated Migratory Bird Treaty between the United States, Canada, Mexico, Japan, and Russia, most recently amended in 1989. Simply put, you can't just go out and shoot the critters without first devoting all your spare time to obtaining permission from an intimidating array of government agencies and then, if you are successful, facing the wrath of the Coalition to Prevent the Destruction of Canada Geese—a goose-rights group that lobbies Congress and issues "action alerts" to apply pressure: "Hilton Hotels Plans to Kill Geese in NJ," and "Help Stop Goose Hunt at Residential Center for Mentally Impaired Adults," and (to show that there is no escaping) "Geese in Alaska Need Your Help!"

Professional assistance, however, is available. On the aggressive end stand the action heroes of commercial animal control—for instance, a New Jersey–based company called A1 Saver, which offers round-the-clock "Emergency Animal Services" and, more menacing, "Wildlife and Bird Solutions." The company handles the necessary paperwork, then goes out and handles squirrels, flying squirrels, raccoons, bats, moles, skunks, chipmunks, rats, mice, snakes, pigeons, opossums, voles, shrews, and, of course, Canada geese. One of its methods is to round up the geese and euthanize them with gas. The company's website displays an image of a person kneeling before a cross, and offers a special tribute to 9/11, which includes photographs of the burning Twin Towers and American warships at sea. In other words, the company is community minded. Nonetheless, because it kills geese, it may still call down the wrath of activists if you engage it to solve your problems.

Retreating through the choices, you come to more special-ized outfits, such as one called Geese Off!, which covers the sub-urbs north and east of New York, including Long Island, and employs border collies to frighten unwanted birds from proper-ties. The border collies don't solve the problem so much as shove it off on someone else. But the company also offers egg-addling services as a more emphatic solution. Egg addling is a form of population control in which federally permitted addlers take eggs from goose nests and either coat them with corn oil, to block the transfer of oxygen through the shell, or shake them vigorously to liquefy the embryos inside. The eggs are then placed back in the nest, because the breeding parents will subsequently continue their care, rather than going out and reproducing again, as they would do if the eggs were simply smashed. The Geese Off! com-pany has a video of addling in action, in which a man with a Brit-ish accent labels and coats five eggs in a nest, while occasionally standing to fend off the angry parents with a canoe paddle. While dipping the eggs in oil, he says contemplatively, "This is a very good nest. This is a good pair, actually, I feel . . ." He hesitates. He says, "I never really enjoy doing this." Who would? But he seems to have been accused of insensitivity in the past.

In the Middle Atlantic states, the self-declared leader in pro-fessional geese-clearing services is a company called Geese Chaser LLC, in Mount Laurel, New Jersey. There, on a fateful day in 1999, a golf course manager with a goose problem spotted a man named Robert Young playing Frisbee with his family's beloved dog, a large border collie they called Boomer. The golf course manager knew something about the breed. He invited Young and

Boomer over to get some exercise by chasing geese. Boomer en-
thusiastically embraced the game. Within a month he had permanently chased away more than five hundred resident geese from
the golf course, with only occasional "tune-up" visits subsequently
required. Young saw a business opportunity, and founded his
company. It grew rapidly as a measure of the popular impatience
with geese. Young outfitted a fleet of cars emblazoned with the
company name and motto—Your Solution to Goose Pollution—
and brought on a large crew. He sold a franchise in Virginia, set
up a goose-chasing school under Boomer's energetic tutelage, and
began to rake in the cash. It was a beautiful business. The dogs
did the work. Young required their handlers to keep them at home
as pets, so the dog rights people were not offended. The geese
were not harmed, so the goose rights people actually supported
the scheme. Furthermore, when the geese flew off to somewhere
else, they created the potential for new customers. Perpetual motion. Boomer was truly a great dog. He died on April 10, 2008. In
his memory, Young commissioned a custom-made motorcycle and
named it after him.

But chasing birds from one property to the next does little for
aviation, and it can be shown that the large numbers of Canada
geese now filling the skies do actually threaten the flying public.
Take their collision record at New York City airports alone. In
June 1995, for example, while landing at Kennedy Airport, a supersonic Concorde absorbed a Canada goose into the number
three engine, which flew apart, throwing shrapnel into the number four engine, destroying it in turn and causing dual-engine
fires. The cost of repairs ran to $9 million, of which the airport
manager, the Port Authority of New York and New Jersey, had to
pay more than half, presumably because it was supposed to
have kept geese away. Three months later, in September 1995, an

Airbus A320 landing at LaGuardia struck more than a dozen Canada geese, including at least one that went into an engine, causing it to torch. The repair cost was $2.5 million. In December 1995, a Boeing 747 inbound to Kennedy flew through a flock of geese, destroying one engine, badly damaging another, and causing extensive damage to the fuselage. The pilot reported that the birds had thumped like sandbags hitting his airplane. The birds in that case turned out to have been migratory snow geese, but this was too fine a distinction. Snow geese are beloved if only for their name, and as usual, Canada geese took much of the blame. The cost of repairs was $6 million. There was a pause for a couple of years. Then, in April 1998, an MD-80 departing from LaGuardia struck Canada geese and suffered serious damage, including a hole punched in the nose. The crew managed to get the airplane across the city to a safe landing at Newark. There were further strikes in May 1998, June 2000, and June 2001. In August 2001, a Boeing 737 hit Canada geese at 10,000 feet while descending toward LaGuardia. Those birds smashed (but did not penetrate) a cockpit windscreen, cutting the captain with broken glass and causing him to depressurize the cabin to keep the windscreen from blowing out completely. There were goose strikes at Newark in October 2002, and again in March 2003. Then, in September 2003, a Fokker 100 lifting off from LaGuardia ran into at least eight of the birds at 150 feet. The airplane was severely battered, and suffered the disintegration of the right engine, which penetrated the fuselage with blade shards and shrapnel, missing passengers only by chance. The crew did a magnificent job. With the airplane vibrating heavily and barely able to fly, they nursed it low across Queens to a safe landing at Kennedy. But it was a very close call. The crew was widely praised. The following month, the pilots of an MD-82 who were descending toward Newark swallowed geese—snow geese again—with both engines, had an explosion on one side, lost control of the airplane, regained control, and landed.

In the New York metropolitan area there were nearly eighty such goose strikes during the decade preceding Sullenberger's encounter. To put that in context, there were more than four thousand strikes with other bird species in New York over the same period. Objective observers, therefore, cannot fault geese alone. The experts at assigning blame are two employees of the U.S. Department of Agriculture, Richard Dolbeer and Sandra Wright, who work out of an office in Sandusky, Ohio, where they preside over the Federal Aviation Administration's National Wildlife Strike Database. The database includes strikes only within the United States, or involving some U.S. airliners abroad. Because it is a repository exclusively for officially filed reports, it is believed to account for merely 20 percent of the strikes that have actually occurred. Nonetheless, for the period from January 2000 to January 2009, it has records of some 98,700 events, involving at least 370 conclusively identified species of birds. The birds have included loons, grebes, pelicans, cormorants, herons, storks, egrets, swans, ducks, vultures, hawks, eagles, cranes, sandpipers, seagulls, pigeons, cuckoos, owls, turkeys, blackbirds, crows, magpies, sparrows, swallows, starlings, chickadees, woodpeckers, hummingbirds, mockingbirds, parrots, and a single parakeet. Over the same period, airplanes officially collided with bats on 253 occasions. Furthermore, they had 763 official collisions with deer, 252 with coyotes, 182 with rabbits, 120 with rodents (including porcupines), 74 with turtles, 59 with opossums, 16 with armadillos, 14 with alligators, 7 with iguanas, 4 with moose, 2 with caribou, and one each with a wild pig and a donkey. There was also an official collision with a fish, though it was in the grasp of an osprey at the time.

The Sandusky database confirms that all of the recorded aircraft collisions with terrestrial mammals have occurred on the ground. The same is true for the aircraft collisions with reptiles. Somehow this is reassuring. Other patterns emerge. Although some birds fly above 20,000 feet, and bird strikes have been reported as high as 32,000 feet in the United States and 37,000 feet in Africa, the density of bird traffic decreases exponentially with altitude. Sixty percent of bird strikes occur within a hundred feet of the ground. That figure reflects the crowds of birds in low-altitude flight, and the basic navigational fact that, when startled, birds on the ground have nowhere to go but up, sometimes directly into airplanes that are landing or taking off. Seventy-three percent of bird strikes occur within 500 feet of the ground. For the 27 percent of bird strikes that occur above 500 feet, Dolbeer has discovered a natural law that he calls the Dolbeer Rule: the number of bird strikes declines by 32 percent for every thousand feet of climb. All told, more than 90 percent of bird strikes occur at less than 3,501 feet. There is variation, however, in the outcome of the strikes. Eighty-six percent of those reported cause no damage at all, in part because so many occur just above the ground, where most birds are small, and impact forces are weaker because airplane speeds there are slow. At the slightly higher altitudes, between 501 and 3,500 feet, the strikes that do occur—some 20 percent of the total—tend to be more dangerous, because airplanes are flying faster and the birds involved are more likely to be large and arrayed in horizontal formation. If you happen to hit one of them, you will likely hit others.

Which brings us back to Sullenberger and Skiles. The database shows that waterfowl are the birds most frequently struck above 500 feet, and that among them the most frequently struck are Canada geese. Aside from their sheer numbers and their year-round presence in busy airspace, no one quite knows why. Dolbeer himself could not explain it to me. He speculated that Canada geese, unlike crows, are simply too dumb or ornery to get out of the way. He didn't bother much to understand their side. The truth is, he seems to have it in for these birds. He respects crows to some degree. He respects ducks, though he hunts them. He made it clear, however, that he does not respect Canada geese. Actually, he made a distinction between the migratory types, about whom he really cannot complain in public, and the implanted full-time residents, against whom, with luck, the encounter over the Bronx would help prepare the way for an all-out campaign.

Such campaigns have long been conducted at LaGuardia itself. They are included in the airport's Wildlife Hazard Mitigation Plan, which is based on a Wildlife Hazard Assessment, and contained in the Airport Certification Manual approved by the FAA. To execute the plan, the airport employs wildlife management personnel, including full-time "bird supervisors," who are trained by "airport wildlife biologists," two of the seven in the United States. The bird supervisors stand watch in relays twenty-four hours a day, restlessly smashing nests and looking for loafing birds, which they disperse, remove, or destroy. They especially have it in for Canada geese. The biggest assembly seems to be on Rikers Island, just off the departure end of Runway 4, where large numbers of the geese insist on nesting, perhaps because they are comfortable with the fact that it is people who are caged there. The choice of location, however, is poor, because every year during their June molting season, when the geese shed their old wing feathers and have difficulty flying, they get raided by officials, who round them up and kill them. The killing is known as culling. It is a joint governmental effort. The raiders include representatives of the U.S. Department of Agriculture, the Port Authority, the New York City Mayor's Office, the New York City Department of Corrections, the New York City Department of Parks, the New York City Department of Environmental Protection, the New York City Economic Development Corporation, the Town of Hempstead, the New York State Department of Environmental Conservation, the U.S. Fish and Wildlife Service, the U.S. National Park Service, and LaGuardia Airport itself. Armed with exemptions and impact statements, these people hit fast and hard.

Dolbeer stands behind them. In four raids from June 2004 to June 2008, his allies culled 1,249 geese from Rikers Island. They cornered them, smashed their nests and eggs, and killed them. But these were stopgap measures at best, because for every "mitigated" goose on the island, some new goose soon arrives to fill the ranks. In LaGuardia's official Master Record, this has to be admitted—if in strangely abbreviated form. "Flocks of birds on & invof apt," the Master Record reports. *Invof* stands for "in the vicinity of." *Apt* stands for "airport." In full translation, the entry reads, "Flocks of birds on and in the vicinity of the airport." Though seagulls were being subjected to an effective "in-house enhanced removal program," for the nominally protected Canada geese, no solution appeared within reach. Among Canada geese, there are perhaps twenty-five thousand full-time residents in the New York metropolitan area, and another twenty thousand migrants that pass through every year. Studies have shown that the residents tend to stay within five miles of home, but around LaGuardia, a five-mile circle includes a large number of ideal nesting grounds, including some, like Central Park, where if you start rounding up geese you will definitely stir up opposition. Killing geese in plain view poses political problems, even if it can be shown to be useful. This was Dolbeer's dilemma.

He has had a successful life. Among his many achievements, he is the second-term chair of the Bird Strike Committee USA. The group is a professional organization consisting of representatives from government agencies, the military, airports, airlines, and aviation manufacturers that serves as a forum for information and technologies and, in conjunction with the Bird Strike Committee Canada, holds annual wildlife mitigation conferences. The 2008 conference was held at a Marriott hotel in Orlando, Florida, despite high winds and rain from a hurricane named Fay. It featured the presentation of more than fifty technical papers and posters.

- *The Florida Statewide Airport Stormwater Survey*. Abdul Hatim, Florida Department of Transportation
- *Let's Talk Turkey*. Christopher Bowser, USDA Wildlife Services
- *Aircraft Bird Strike Avoidance Radar Systems—Looking Forward*. T. Adam Kelly, DeTect, Inc.
- *Teamwork in Large Countries Like Brazil—A Good Solution for Bird Strikes*. Flavio Antonio Coimbra Mendonca, Brazilian Air Force
- *Rapid Dispersal and Long-Term Effect on Canada Geese by On-Demand Alarm and Alert Call Playback Reinforced with Cracker/Banger Shells*. Phillip C. Whitford, Capital University

And so forth. These presentations were open to the interested public. Some of the papers later appeared in the journal *Human-*

Wildlife Conflicts, which is published by Utah State and Mississippi State universities. Dolbeer offered the welcoming and closing remarks. In the audience were 450 attendees who had flown in from all over the world. It was a record turnout, and clearly a sign of the times.

In that sense Sullenberger's accident was an opportune event. The Sandusky database recorded it as if it were just another strike, but Dolbeer knew better. With all the publicity and the talk of heroes, this was an opportunity to bring the public into line, and launch a serious program of eradication. Dolbeer was on edge during the NTSB's investigation, with which he was closely involved. Given the position and direction of flight of these particular geese, it seemed unlikely that they had come from within five miles of the airport—and that was unfortunate—but it would certainly be useful if the analysis of their remains would show them at least to have been of the resident-goose variety. Dolbeer and other investigators extracted blood, feathers, and other organic matter from the ruins of the Airbus, and sent sixty-eight samples to the Smithsonian's Feather Identification Lab. The lab is a bird strike forensics facility near the NTSB in Washington, D.C. It is led by a woman named Carla Dove, who is renowned for her expertise. After several weeks spent studying the samples, she sent over the initial findings: on various occasions in the past, the airplane had collided with a songbird, a pigeon, and a duck, but the rest of the samples collected had come from Canada geese.

So far so good. But it was unknown as yet whether they had been migratory or resident birds, apparently because the two types are so closely related that the distinction could not be made with DNA. Dove required another few weeks while she subjected the feathers to a comparative stable hydrogen isotope analysis, allowing insight into the birds' diet at the time that the feathers formed. At last came the news: the geese involved in the accident had not grown their feathers near New York, but in remote areas of northern Canada, either in distant Newfoundland or in the wilderness around Hudson Bay. Simply put, the birds on January 15 were legitimately flying by as their ancestors had since before New York was settled, and no amount of local egg addling or goose gassing could have kept this accident from happening.

The anti-goose forces staggered from the blow, but quickly rebounded: upon further consideration, it turned out that the distinction between migrant and resident geese is not so important after all. What matters is the numbers of these birds collectively, and the effect that any of them may have on an airliner that collides with them. What better evidence is required than the US Airways accident? Dolbeer in due course appeared at the NTSB hearing, and this was the subtext of his testimony. He is a careful and knowledgeable man, America's preeminent expert in bird encounters, and his opinion is hard to ignore. It cannot be by chance that immediately after he made his case in public—and indeed, on June 11, the last day of the hearing—New York City, the Port Authority, and the U.S. Department of Agriculture announced plans for an all-out effort to round up a record two thousand resident Canada geese on parks and properties within a five-mile radius of both LaGuardia and Kennedy airports. The molting season was on, and as Dolbeer believed, this was the moment politically to crank up the pressure.

Word spread quickly, and on Tuesday, June 16, a group of protesters gathered in front of the Port Authority's Manhattan headquarters. One of the protesters was a thin, tall pilot who held up a sign reading "Pilot Against Goose Gassing." Others held up signs objecting to the unusually high toll. They chanted, "How many geese did you gas today?" The spokeswoman for the group explained to the press that the resident geese being targeted had been shown to be innocent. She was Edita Birnkrant, director of an organization called Friends of Animals. Speaking of the pedestrians and drivers passing by, she said, "A lot of people are giving the thumbs-up and supporting us. Some people don't even know this is going on." But it was too late. Though the protest made *The New York Times*, both the Port Authority and City Hall easily dismissed it. A Port Authority spokesman said, "Our responsibility is to think about safety for people before peace for geese." If Birnkrant did not agree, well, she was a known figure in the New York scene. Just the week before, she had been active in a protest to ban the horse-drawn carriage trade, and in that, too, had been ignored. Now, she could protest as loudly as she pleased—that was her right as a New Yorker—but already the new program was under way, and the first hundred geese had been culled.

THREE

THE COLLISION

3:27 p.m.

Sullenberger and Skiles, by contrast, were not out to kill geese. They expected to finish their flight and go home. But then Sullenberger said, "Birds!" and Skiles said, "Whoa!" and the geese seemed to fill the windscreen just outside. Again, the encounter was roughly head-on, with Flight 1549 climbing toward the north, and the birds cruising toward the southwest. The closing speed, therefore, was a combination of the airplane's 250 miles per hour and an angular component of the geese's own forward speed, which may have added 25 miles per hour to the sum. At that rate, any geese suddenly filling the view should have slammed into the windscreen, if not gone through the glass, and Sullenberger should indeed have ducked below the glare shield to protect his head. He did not, and they did not, but at least five geese hit the airplane anyway—one on the underside of the nose, one just below the cockpit's right side window, one on the left wing, two or more into the left engine, and at least one into the right engine. The thumps were clearly heard in the cockpit and cabin, followed by a shuddering sound.

According to the official transcript, Skiles swore. He said, "Oh #!"

Sullenberger said, "Oh yeah." It sounded to him as if the sky were raining birds.

The engines banged with compressor stalls—small explosions that are similar to backfires and usually harmless in themselves, but may be symptoms of serious underlying problems. In this case the underlying problem was that both engines had just been trashed. Jet engines are air compressors. They gulp the outside air, compress it with fans and fire, and shove it out the back at high speed. By shoving air backward, they propel the airplane forward. More accurately, they impart energy to the airplane that can then be spent on combinations of altitude and speed. Because they are inherently simple devices and built of the finest materials, they very rarely fail. But in this case, within three seconds of the impact, they began to wind down. The loss of thrust was shocking.

Skiles said, "Uh-oh."

Sullenberger said, "We got one . . . Both of 'em rolling back." He had never experienced an engine failure before. Now he was experiencing two.

In the cabin, the veteran flight attendant Doreen Welsh was sitting in the aft galley strapped into a forward-facing jump seat with a view up the aisle toward the front. The other two flight attendants, Donna Dent and Sheila Dail, were sitting side by side just behind the cockpit, facing aft. They felt the thumps and heard the engines wind down. Dail whispered, "What was that?" Dent answered, "Probably a bird strike." The cabin turned eerily silent. An engine slowly clanked. The cabin filled with a trace of smoke, accompanied by a burning smell. Dent told Dail that this, too, could have been caused by a bird strike. Up in the cockpit, Sullenberger assumed that the scent was coming through the pressurization ducts, and that it was the smell of burning birds. In the cabin the flight attendants thought it resulted perhaps from short-ing wires. Passengers behind the wing saw large flames trailing from the left engine, and concluded that the engine was on fire. It was not. Unburned fuel was passing through the crippled com-bustion chamber and torching harmlessly in the slipstream. The flight attendants did not see the flames. From their seats, they could not see outside. They assumed that the airplane would be returning to LaGuardia. In the back of the cabin, Welsh un-strapped herself from her seat and went forward quickly, check-ing the overhead bins for the source of the smoke. She was heard to say something about a fire extinguisher. A woman sitting in the last row, on the aisle, stood up and may have tried to retrieve something from overhead; this is not clear. She later said that she offered to help, and asked Welsh if she should call a flight atten-dant for her. According to her, Welsh snapped, "I *am* a flight at-tendant!" and the woman snapped back, "I meant *another* flight attendant." Welsh ordered her to sit down. She did. The confron-tation was noticed by others. Welsh returned to her jump seat and strapped herself in again. She assured the passengers in the back that everything would be fine. She tried to raise Dail over the in-tercom phone, but somehow they could not make it work. As of yet there had been no word from the cockpit.

The pilots were busy. When the airplane hit the birds, it was climbing with its nose pitched up 10 degrees above the horizon. As the engines wound down, the deceleration was dramatic, in part because for about fifteen seconds the airplane stayed nose high and continued to climb. Skiles still had the controls at that time. Sullenberger urgently tried to restore thrust to the engines. They were still turning, but at very low speed. It was possible that they had simply flamed out, and that with the standard engine-start igniters he could relight the fires. He said, "Ignition start,"

and rotated a knob one click to that position. The igniters began
to click, but the engines refused to respond. They simply were not
meant to swallow geese and survive.

.

No engines are. Not that bird strikes are ignored in the design of
jet engines. Indeed, engines are built and certified with the possibility of bird strikes in mind. The regulations are complex, and
they have become gradually more stringent over the years. For
these purposes, bird species are officially categorized as small,
medium, and large. The small ones weigh less than 3.3 ounces.
Because of the density of their flocks, they are the birds most
likely to strike multiple engines, and each engine multiple times.
Engine manufacturers have to demonstrate that their designs will
continue to produce climb thrust even after being hit by a group
of them—one little bird for every forty-nine square inches of engine inlet, up to sixteen little birds in rapid succession. The demonstrations are carried out on engines attached to rigid stands and
spooled up to climb thrust. The birds are commercial, farm-raised
stock purchased from suppliers. They are slaughtered just before
the tests, then wrapped in lightweight Styrofoam sabots, loaded
into nitrogen-powered pneumatic cannons, and fired into the engines at about 250 miles per hour. The cannons are known vari-
ously as chicken guns, turkey guns, or rooster boosters. The tests
are filmed with high-speed cameras and can be viewed on the
Internet in slow-motion videos, some set to music. In real time,
the birds pass almost instantaneously through the test engines.
They go in whole and come out as spray. Animal rights advocates
have objected to this. A researcher in England is trying to accommodate their concerns by creating an artificial standard-density
bird—a Jell-O bird, it is called—that will spare the test birds for

some other fate. This turns out to be difficult to do, because real birds, though gelatinous, have bones, muscles, and sinews. Indeed, there is a concern among some regulators that the farm-raised test birds being used today are themselves unrealistic because they are flabby compared to their wild brethren, who seem to cause more damage for birds of the same weight.

It's a niche worry, and should be optional for the traveling public. In practice, the industry has come a long way in producing engines that can swallow small birds, and even medium-size ones (such as seagulls; officially, up to 2.5 pounds) without disintegrating or losing significant thrust. The reasons are not difficult to understand. Modern airline engines are hybrids, called turbofans, each of which contains an old-fashioned jet engine in its core, but develops most of its thrust not by shooting a column of high-speed exhaust out the back (as in pure-jet designs), but by reaching forward through itself with a central shaft and driving a propulsion fan. That fan is what you see when you look into the front of an engine. On the engines that powered Sullenberger's Airbus, the fan has a six-foot diameter. It is really just an air pump, similar to an ordinary window fan, but many-bladed, jet-powered, and enormously more forceful. Even when throttled back to minimum speed on the ground, it is capable of sucking in airport workers who stray closer than about six feet to the inlet. More usefully, when it is throttled up to takeoff, climb, or cruise settings, it ingests huge masses of outside air, which it accelerates rearward through the engine casing. At the center of the engine, and just behind the fan, a portion of the accelerated air feeds directly into the jet core, where it is compressed, burned in kerosene-fueled fires, and used to spin turbines (primarily to power the compressors and fan) before being shot as a hot gas out the back. Far more of the fan's accelerated air, however, completely bypasses the jet core and rushes unheated to the rear of the engine, where it returns to the atmosphere. The blown air is known as bypass air. On the A320, it provides as much as 80 percent of the engine thrust.

The fan, in other words, is the ultimate focus of jet-engine design. Its blades overlap, are set at a sharp angle, and are made of strong, flexible, lightweight titanium. These are what birds first hit on the way in, and for the birds the encounter is traumatic. In fact, the birds are liquefied. The effect varies little according to size. Small, medium, or large, the birds become an instant soup— a bloody sludge that is known in the business as slurry, and is said to leave engines with a telltale smell, enhanced with fishiness after the liquefaction of fish-eating fowl. This is the sort of knowledge that investigators acquire only in the field, after airplanes have crashed or serious bird strikes have been reported. It requires dedication to discovering truth in such cases, and a certain investment in the idea that accuracy matters.

In any case, turbofan engines are self-protective to some extent, because, when hit by birds, the fan blades may bend without breaking and sling the bird slurry outward, forcing it to blow harmlessly through the bypass ducts, perhaps splattering against protrusions, but never entering the power source—the critical high-speed components that constitute the jet core. And this is not just brave talk. The Sandusky database indicates that of the 12,028 engines reported to have been struck by birds between 1990 and 2007, two-thirds emerged unscathed from the encounters. Of the remaining third—the engines reported as damaged— more than 90 percent continued to produce thrust in some manner, and only 312 were totally destroyed in flight. In short, complete engine failures following bird strikes are rare.

Some, however, will inevitably occur. The reason is that, within the constraints of materials science and practical design, it is simply not yet possible to build turbofan engines that can reliably withstand 250-mile-per-hour collisions with even single birds heavier than the official medium size of 2.5 pounds. In recognition of these realities, certification requirements for the official big-bird test do not require the engine to keep producing thrust, but merely to accommodate its own destruction without running angrily out control, throwing dangerous shrapnel through the engine casing, or bursting fuel lines and catching on fire. Currently, the weight of the big birds used is eight pounds. That is lighter than millions of birds flying around in the North American skies, including typical twelve-pound Canada geese, but it is heavy enough to ensure the death of the (very expensive) test engines. The big-bird tests are single shots, aimed toward the center of the fan, to ensure that parts of the bird are ingested into the critical jet core. Usually a chicken is volunteered for the job. The destruction starts when the bird hits the fan. Even as the bird is turning into slurry, it causes fan blades to bend, erode, and fracture—reducing the fan's thrust and sending a hail of titanium debris deeper into the engine. Some of the debris exits harmlessly with the bypass air, but other debris (now mixed with slurry) finds its way into the spinning compressors at the entrance to the jet core, where it sets off a cascade of successive failures, with shattered blades and vanes adding to the destructive hail. In response to the disruption, temperatures inside the combustion chambers may rise so high that the debris passing through is turned to molten metal, which splatters against the downstream turbines, even as they themselves are being warped and destroyed by the heat. Needless to say, any part of the bird that has made it this far is vaporized. Meanwhile, overall, the engine will likely be convulsing as it dies.

This was approximately the situation faced by Sullenberger and Skiles. Their engines were CFM56-5Bs, built in Ohio by a

General Electric–Snecma consortium and considered to be among the world's most successful designs, but certified for use at a time when the regulations required a big-bird test of only four pounds. Considering that the Canada geese the engines swallowed on January 15 probably weighed three times as much, and that multiple birds were involved, the engines behaved extraordinarily well simply by not exploding. The right engine had taken the worse hit. Its jet core was now barely functioning, and its propulsion fan was spinning at less than 17 percent of its designed maximum speed—a useless sub-idle setting. The left engine was more valiant. Though its fan speed had dropped to merely 36 percent, and was producing very little thrust, its jet core was still functioning well, continuing to spin at an impressive 83 percent of its maximum speed. That left-engine core was central to the flight, because it continued to power the airplane's hydraulic and electrical systems—without any degradation of the computers or controls.

Sullenberger and Skiles were both aware that the left engine was to some degree alive, but this was hardly the time to marvel over its resilience or pause to admire its contribution. The situation was critical. They were at low altitude over New York City, pointed north away from the airport, and condemned to a descent with no appreciable thrust. Sullenberger at that moment showed a masterly presence of mind. Rather than falling back on by-the-book procedures, he improvised in a way not covered in his training, by reaching to the overhead panel and starting the auxiliary power unit—a small turbine engine in the tail that can be used for main engine restarts, and also drives a generator that would provide for full electrical power should the left engine completely fail. Electrical power is essential for the A320's flight control system. Sullenberger kept Skiles informed. He said, "I'm starting the

APU." Then he did his duty as pilot in command, saying, "My aircraft," and took his control stick in hand.

Skiles released the control stick on his side. He said, "Your aircraft."

Sullenberger said. "Get the QRH." QRH stands for Quick Reference Handbook. He said, "Loss of thrust on both engines." Skiles already had it out. It was a checklist for the engine-restart procedure.

At New York Approach, the controller Patrick Harten had no notion of the trouble they were in. Cactus 1549 still looked normal on his radar screen, a routine Airbus flying north and climbing. Since his last transmission to the crew he had been working other traffic. Now he was going to turn the flight west and start moving it toward its destination. He radioed, "Cactus 1549, turn left, heading two-seven-zero [degrees]."

Sullenberger was rolling into a left turn already, but to get back to the airport. While lowering the nose to maintain a good gliding speed, he radioed, "Mayday, Mayday, Mayday. Ah, this is, ah, Cactus 1539, hit birds." Under stress, he had thrown the call sign by a digit. So what. He said, "We lost thrust in both engines. We're turning back towards LaGuardia."

It was a shock, but Harten was up for it. Without the slightest hesitation, he responded, "Okay, yeah, you need to return to LaGuardia. Turn left heading of, uh, 220." A heading of 220 degrees means southwest.

Sullenberger said, "Two-two-zero." He was completely concentrated. He had prepared for this moment all his life, he believed.

FOUR

THE GLIDE

3:28 p.m.

The airplane did not plummet toward the earth, as headlines later suggested. Indeed, it descended rather gently. Sullenberger nosed into the glide from a position about 3,000 feet above the city. From that moment until the landing in the Hudson, three minutes and twenty-one seconds passed. The average descent rate, in other words, was something less than a thousand feet a minute. There is nothing surprising about this. Sullenberger flew close to the optimal speeds. There was some thrust coming from the left engine—not much, and not throughout the glide, but it helped. More important, the landing gear and flaps were up, so the airplane was aerodynamically clean. The gentleness of descent was a function of the airplane's design. And it goes a long way toward explaining why the passengers remained calm.

With a near-total loss of thrust, some other airplanes would have descended more rapidly, to be sure. The old F-4 Phantom comes to mind. It was the heavy supersonic fighter flown by American forces in Vietnam and beyond—a twin-engine, two-crew, tandem-seated brute that seemed to rely entirely on muscle to stay in the air, with mere afterthoughts for its wings and its downturned tail. Aerodynamically, it was actually an excellent

design for the 1960s, but compared to fighters since then, it was
something of a pig. This was the airplane Sullenberger flew in the
Air Force in the late 1970s. He flew it largely in Nevada, never in
combat. Recently he told me that it gave him the best flying of his
life, low and fast across the wastelands, or high, very high, and
even faster. You can't fly an Airbus that way, but that's all right,
because there are different times for different sorts of flying in
any pilot's life.

It is obvious that Sullenberger was always a superb pilot, however the definition has changed for him across time and however
many other professional pilots share the distinction. Like others,
he was a pilot's pilot, not greedy for power or wealth, a steady and
decent man, but tied to a profession in decline. He grew up as the
son of a dentist in Denison, Texas, with an air base nearby. The
airplanes passing overhead inspired him. He learned to fly in
high school, on a local grass runway in an ancient Aeronca without radios, electrical system, or starter. Because it had small performance reserves, it taught him a lot. He soloed at age sixteen,
earned a private license at seventeen, and went off to the Air
Force Academy, where he was one of the few licensed pilots
among the cadets. Flying became the defining passion in his life.
Now, at age fifty-eight, he still grows tender about the grace of
aerial motion and the fluidity of control. Beneath the jargon of
professionalism, this is what he really knows. He speaks about it
as pilots do: sweeping his hand into a carving bank, or a flat forward acceleration, or, the sure giveaway, a mushing palm-down,
fingers-raised descent.

"Mushing" is a term used to describe slow flight, when the
nose is pointed up though the airplane may be descending, and
the wings are plowing through the air at a steeper than usual
angle. A mushing palm-down, fingers-raised descent is the most

beautiful expression of wings. It is why pure landings are done with no flaps or leading-edge slats, allowing the wings to speak entirely for themselves. It is also how a Phantom flies after both engines have quit and the airplane has been trimmed up to the recommended 250 miles per hour for the glide. In that configuration, without thrust, the Phantom loses at least 3,000 feet per minute—a high rate at which to be closing on the ground. There are other problems with dual-engine failure in that fighter, including the loss of the hydraulic power that is necessary to maintain control if the airplane is slowed below speeds that keep the turbines briskly windmilling. Nonetheless, though the situation is difficult, it is not impossible. In theory, a perfect pilot could shoot across a runway 10,000 feet high in his flamed-out Phantom, then bring the airplane around through a series of steep banks to line up on a well-assessed final approach, then make a few little S-turns, like switchbacks, to tweak the angle, then temporarily slow the descent rate by raising the nose on short final approach into a pre-flare, a descending zoom similar to the one the Space Shuttle performs, and then, just above the runway, flare a second time, maintaining sufficient speed for the hydraulic controls, but trading some of that speed for descent-stopping energy to keep from driving the landing gear through the wings on touchdown. A perfect pilot with perfect luck on a perfect day could do it. It would be a glorious occasion. Almost certainly the pilot would then acknowledge the airplane to some degree, thinking, Thank you, Betsy, and probably saying it aloud. Certainly no perfect pilot would walk away without looking back. But procedures in flight are designed for average pilots with no luck involved. In the case of dual-engine failure, F-4 Phantom crews were given no chance to be creative. They were required without exception to eject from the cockpit—to punch out and let the airplane go to hell.

Sullenberger never had to. For five years he flew Phantoms without drama, until, in 1980, he left the Air Force and joined the airlines, where punching out was no longer possible but the airplanes naturally provided for more gradual descents to the ground. Indeed, in the pursuit of efficiency, jet airliners over time have become so good at staying aloft with minimal power that workaday descents require significant advance planning and often cannot be performed as decisively as air traffic control desires, particularly when speed reductions are also required. Sullenberger makes the parallel point: that the latest generations of airliners have grown significantly cheaper to fly per passenger mile. This is due largely to improvements in aerodynamics and, most fundamentally, to the introduction of long, thin wings meant to lift well at high altitude and with minimal drag. The purpose is to milk the maximum range from the fuel aboard. The engines are not quite afterthoughts—and important gains have been made there, too—but to a striking degree, airliners across the generations have come to resemble high-performance gliders, or "sailplanes," the ultimate in efficient flying machines. Sullenberger flew gliders as a cadet at the Air Force Academy, and he worked as a glider instructor for a few summers before becoming a fighter pilot, but the experience was less useful to the emergency facing him now than it might appear to have been. Sullenberger said it himself to the NTSB: he had not flown gliders for years. Instead, he credited his success to basic knowledge of "energy management," a grand term for the routine trade-offs between speed and altitude that are frequently made on every flight. Also relevant to the success were his decades of flying airliners that are good gliders in their own right, and prove it daily during routine descents with passengers aboard. During those descents the engines may be throttled back to a minimum setting known as "flight idle," at which they produce hardly any thrust at all and, unbeknownst to the passengers, the airplanes glide for as much as fifty miles until arriving at the desired lower altitudes, where power is again applied.

Of course, the mark of a true glider is that it has no engine at all, and therefore has no power to apply at the end of a descent. The solution in high-performance sailplanes is to find atmospheric lift, and ride the rising currents to gain altitude and stay aloft. Because such sailplanes are capable of losing as little as one hundred feet per minute, the merest lift suffices. It is routine after an initial tow to a low-altitude release position to fly full days and hundreds of miles before coming in for a landing. Such are the capabilities of these specialized aircraft that sailplane endurance attempts were canceled after a Frenchman stayed aloft for fifty-six hours in 1952, and the official record keepers decided that this was getting to be ridiculous and a threat to pilots' health. The current distance record for sailplanes is 1,870 miles, flown in Argentina in 2003 by a German pilot named Klaus Ohlmann, who exploited atmospheric lift generated by winds blowing across the Andes. The corresponding altitude record is 50,722 feet, about 10,000 feet higher than airliners normally cruise. It was set in 2006, also in Argentina in Andean lift, by two American pilots wearing pressure suits—the adventurer Steve Fossett (who died a year later in a small airplane in California) and a former NASA test pilot named Einar Enevoldson, who specialized in ultrahigh flying. Again, this was in a glider.

•

It is obvious that no one will set soaring records in an airliner without power, but history shows that a total loss of thrust is not necessarily catastrophic. There was the 1982 case, for instance, of a British Airways Boeing 747 that flew through a volcanic plume one night over Indonesia and suffered compressor stalls and the loss of all four engines at 37,000 feet. The ensuing glide (with engines harmlessly belching fire) was written up afterward as a "near-death experience" for the passengers, during which the airplane "plunged." But "near-death" is a relative concept, and "plunging" in no way describes what actually occurred. In fact, the crew had nearly twenty minutes of available gliding time, during which they figured they could reach a certain airport about a hundred miles distant, so long as they had sufficient altitude to clear a mountain range that stood in the way. The pilots were hardly relaxed. They were making Mayday calls to Jakarta Control, flying the airplane, handling the depressurization of the cabin, and struggling with procedures to restart the engines. Nonetheless, in the midst of the glide, and with appropriate British aplomb, the captain announced to the cabin, "Good evening, ladies and gentlemen. This is your captain speaking. We have a small problem. All four engines have stopped. We are doing our damnedest to get them going again. I trust you are not in too much distress." The captain's name was Eric Moody, to give credit where it is due. A few people were indeed in distress and spent the ride sobbing or whimpering, but most seem to have matched the captain's calm. A British spinster traveling with her aged mother is known to have turned back to reading a Jane Austen novel. Apparently she just was not going to stand for this nonsense. It turned out that her attitude was right. As the airplane descended below 12,000 feet, the crew managed to restart the engines.

The following year, on a Saturday afternoon in the summer of 1983, an Air Canada flight ran out of fuel at 41,000 feet above the plains of Manitoba. It was a twin-engine wide-bodied Boeing 767 on a domestic run from Montreal to Edmonton, with sixty-nine people aboard. The 767 was a new design at the time, and the first airplane in Air Canada's fleet to use metric rather than imperial measures. As a result of related confusions compounded by instrument failures and faulty procedures, the crew embarked on the trip carrying only about half the fuel they required. The captain was an experienced sailplane pilot named Robert Pearson. After low fuel-pressure indications suddenly alerted him to the condition they were facing, he began a diversionary descent toward the nearest major airport, in Winnipeg. At that point the engines were still running. But as the airplane descended through 28,000 feet, both engines quit—first the left, then the right. Total engine failure was not a contingency that Air Canada crews had trained for, or that was covered by Boeing in the 767 operations manuals at the time, but Pearson did what one would expect of any professional pilot, and trimmed the nose up to what he estimated (correctly) to be the optimal gliding speed of 250 miles per hour. In the cabin the passengers remained calm, though perhaps less in a proud British style than in the Canadian tradition of placidity. The pilots were calm, too, even though, heck, it sure looked like they weren't going to make Winnipeg anymore.

As luck would have it, a decommissioned air base lay only about fifteen miles away, near the small town of Gimli, Manitoba. The air base had been built with parallel runways, one more than 7,000 feet long. Pearson spotted it ahead and lined up on the runway for a straight-in approach. He realized that for lack of hydraulic pressure he would be unable to operate the wings' flaps, slats, or speed brakes, and that he would have to rely on a problematic free-fall mechanism to lower the landing gear. Sure enough, when he tried to lower gear, the nosewheel refused to lock firmly into place. A larger problem was that the airplane was coming in too high and fast, and was going to overshoot the runway and crash in rugged terrain on the far side. Unable to brake by conventional means, Pearson performed a maneuver from the biplane era never practiced on jet airliners, though sometimes still used in sailplanes: he applied full right rudder and banked hard to the left, putting the giant swept-wing 767 into a cross-controlled "slip" and creating aerodynamic drag by forcing the fuselage against the airflow. It is a testament not only to Pearson's skill but also to the grace of Boeing's design that he was able to get away with this at low altitude, on a short final approach. Thank you, Betsy. Around this time, mere seconds before touchdown, the pilots suddenly realized that the runway was no longer actually a runway, and that it had been converted into a drag strip with a guardrail running down the center line. To make matters worse, an event was being held there, with cars parked along the sides and a crowd of people having a barbecue at the far end. Some people looked up, saw the jet coming, and began to run—but most remained unaware.

Pearson had done a good piece of flying. Call it energy management, if you will. He touched down hard merely 800 feet past the old runway threshold, and laid heavily into the brakes in an all-out attempt to stop the Boeing before it plowed into the crowd. Several main tires blew with loud explosions, the nose gear collapsed, and the airplane went sliding down the pavement on its chin, throwing sparks and chasing people to both sides before coming to a stop. No one was seriously hurt, and surprisingly little damage was done. After only two days the airplane was patched up, fueled up, and flown away. Known afterward as the Gimli Glider, it remained in service with Air Canada for another twenty-five years, until, in 2008, it was retired with fanfare to rest among other worn-out airplanes at the desert airport in Mojave, California. Pearson and his copilot rode along on that last flight to honor the machine. They had initially been demoted by Air Canada for running out of fuel, but after the union protested on their behalf, they were rehabilitated as heroes and they went on with their careers. Two years after the incident, in 1985, they were given an airmanship award by the Fédération Aéronautique Internationale, a fussy Switzerland-based organization that presides over world records.

.

Engines-out airliner gliding is not a sanctioned category, but records exist nonetheless. The current holder is another Canadian, named Robert Piché, who, if not exactly a rebel, seems to be something of a maverick. In August 2001 he was flying as the captain of a wide-bodied Airbus A330 at night over the Atlantic, and was a thousand miles from Europe when he ran out of fuel and lost thrust in both engines. The airplane belonged to a large Canadian charter company called Air Transat, and was carrying 306 people overnight from Toronto to Lisbon. More than half of the passengers were Portuguese. Piché was a French Canadian par excellence—a man roughly of Sullenberger's age, but with the soul of a buccaneer. He had grown up on Quebec's remote Gaspé Peninsula, in a town with a locally significant airport, and, like Sullenberger, had learned to fly as a teenager. While Sullenberger had gone off to the Air Force and a regimented airline career, Piché had moved in a different direction, becoming a bush pilot and flying every manner of old airplane on every manner of mission, including a run in 1983 from the Caribbean to Georgia, in a Piper Aztec, bringing in a load of marijuana. Disapprove if you like, but a run like that took particular courage because it was so

completely solo and risky. Piché got caught. He served sixteen
months of a ten-year sentence in a Georgia prison. Afterward, he
returned to Canada and a checkered flying career, until in 1996,
at the age of forty-three, he managed to hire on with Air Transat,
an outfit with the decency not to hold his transgression against
him. He rose rapidly from copilot to captain on Lockheed L-1011s,
and transitioned to the Airbus A330 in the spring of 2000, after
four weeks of simulator training at the factory in France. He liked
the airplane—who would not? It was another miracle of engineering, a large-size derivative of the A320, and similar in its design
philosophy. But now, just over a year later and far out over the
Atlantic, his airplane sprang a leak and began to lose fuel—first
from the right tanks, where the leak had occurred, and then, because of a valve that Piché mistakenly opened, from the left tanks
as well.

It was 5:36 in the morning local time. They were at 39,000
feet on a clear black night, with stars overhead but nothing more
in sight. Nine minutes after the initial warning, the quantity indi-
cators showed insufficient fuel to reach Lisbon, and Piché decided to backtrack to the nearest airport, on Terceira Island in
the Azores, about 200 miles to the southwest. He made the turn,
and the copilot advised Oceanic Control of the situation. It was
5:48 a.m. The senior flight attendant entered the cockpit to discuss the passenger services that would be required in Lisbon. Piché advised her of the low-fuel condition and of the precautionary
diversion to the Azores. She left to inform the cabin crew and have
the snack trays picked up and the galleys secured. Piché had asked
her to see if she could spot a fuel leak in the form of a vapor trail
above or behind the wings. She tried—even dimming the cabin
lights to improve her outside vision—but saw nothing, and re-

turned to the cockpit to report as much. Piché now told her to prepare for a ditching. A ditching is an intentional water landing. In a jet airliner at night in the open Atlantic it means near-certain death for everyone aboard, no matter how strong the airplane is or who is flying it. The flight attendant went back and calmly instructed the cabin crew to prepare some life vests for passenger demonstrations.

It was 6:06 a.m., a half hour since the first sign of trouble, and still pitch-black outside. The cross-feed valve remained open, continuing to drain the left tanks into the leak on the right side. Seven minutes later, at 6:13 a.m., the right engine quietly flamed out and the airplane began to drift down from 39,000 feet, still 170 miles from the airport. The passengers would not have been aware of anything unusual had not the cabin crew suddenly switched on the lights, taken up positions in the aisles, and begun instructing people to put on the life vests that were stored under the seats. This is not a pleasant way to be woken up on a transatlantic flight, and the mood was not exactly calm, but the passengers had the wherewithal at least to get their vests on. In English, French, and Portuguese, the crew proceeded with the standard briefings for impact-bracing, ditching, and an evacuation into the water.

As the airplane drifted through 37,000 feet, Piché spotted the lights of the island about 140 miles ahead. But at 6:26 a.m., thirteen minutes into the single-engine descent, the left engine used up the last of the fuel, and the A330 became a glider. Piché started swearing in French. *Putain*. They were 90 miles from the airport, slowed to the best gliding speed, and descending through 34,500 feet at about 1,200 feet per minute. From the underside of the airplane an emergency windmill known as a ram air turbine automatically extended into the slipstream and began spinning to provide backup hydraulic power and the minimum of electricity. In the cabin, the regular lights flickered and went out, and the dim emergency lighting came on. This did not go over well with the passengers, some of whom began to cry and pray out loud. To make matters worse, the public address system failed. Normally the failure would have been a blessing, but it was inconvenient now. Five minutes later, as the cabin pressure leaked away, the oxygen masks automatically dropped from the ceiling, and this caused another round of consternation. Up in the cockpit, the situation was more sober, but also quite rough. Piché and his co-pilot were so busy that they did not put on their oxygen masks. They had lost most of the airplane's electronics and flat-panel displays, and were flying with degraded controls, which offered little of the assistance normally provided by the A330 to its crews. Later, Piché implied that his entire life had led up to this moment in flight—a tautology that seems to have become the standard claim in these cases. He even included prison in the progression, and credited it for teaching him not to shy away from realities, however grim.

Not that he had a choice. At 6:31 a.m., while gliding down through the night, 27,000 feet above the ocean and thirty-three miles from the airport, he checked in with Approach Control and requested that the runway lights be flashed. At 6:39 a.m. he arrived nine miles off the approach end of the runway and 13,000 feet high. Being high allowed for plenty of maneuvering for flight path control. It introduced room for piloting errors but also for piloting skills, and took away the element of chance. Piché swept the airplane into a descending 360-degree full-circle turn during which he extended the leading-edge slats and lowered and locked the landing gear. He straightened out from the turn at 8,000 feet on an extended final approach. The runway ahead was 10,866 feet long. It was outlined in lights. Piché saw that he was high, and put the Airbus through a series of S-turns much as some F-4 Phantom pilots might have wanted to do when flamed out, if only it had been allowed. Piché was flying at F-4 speeds or faster, though with descent rates much lower. In the cabin the flight attendants were screaming, "Brace! Brace! Brace!" to the terrified passengers. The airplane crossed the runway threshold doing 230 miles per hour, about 50 miles per hour faster than a normal full-flap speed. It slammed against the pavement about a thousand feet along, bounced back into the air, and floated for another 1,770 feet until Piché goddamn planted that airplane down to stay and locked the goddamned brakes. The planting did not drive the landing gear through the wings, but it was violent enough to wrinkle the fuselage. The locked tires slid for about 400 feet, then abraded and deflated, leaving the airplane to grind to a halt on the ruins of its wheels. It stopped 2,400 feet from the far end of the runway. It was 6:46 a.m., at the end of a world-record 20-minute, 34,500-foot, 90-mile, 306-person, engines-out airliner glide. The passengers evacuated down the slides. Piché followed, and walked around the airplane. It would be repairable, but meanwhile was a wreck. Piché returned to a hero's welcome in Canada, where at first he ducked the publicity, then learned to enjoy it and seek it out. He accepted multiple awards, authorized an official biography entitled *Robert Piché: Hands On Destiny* (available in French or English, autographed), built an official Captain Robert Piché website to which you may provide your email address ("Captain Piché activities: be informed!"), and set himself up as an inspirational speaker for business groups (teamwork and resolve) and schools (excitement about aviation). It turned out that Captain Piché was quite a speaker. He returned to flying, however, because, *putain*, piloting is what he knew best.

•

Captain Sullenberger, by contrast, hardly ever swears. Furthermore, the closer he gets to airplanes, the more straitlaced he becomes. You can hear it in his tone all through his transmissions to New York Approach. He was calm, concentrated, and appropriate. Skiles was equally steady. As Sullenberger put the airplane into a precise 33-degree banked turn, Skiles called out the items from the quick-reference checklist. Skiles said, "If fuel remaining, engine mode selector, ignition, ignition."

Sullenberger had already done it. He said, "Ignition."

Skiles said, "Thrust levers, confirm idle."

Sullenberger had his right hand on the throttles. He pulled them back to the position for flight idle. He said, "Idle."

The right-engine fan, which was already spinning at sub-idle speeds, did not respond. The left-engine fan, which had been spinning at slightly greater speeds, slowed to idle.

Skiles said, "Airspeed, optimum relight. Three hundred knots. We don't have that."

Sullenberger agreed. "We don't."

That part of the checklist was written for dual-engine failure at high altitude, where there would be space to lower the nose and accept a high descent rate in order to gain speed for a windmilling start. This was not an option for them now, during a turn at low altitude with New York City just beneath. Other options, however, existed. Skiles kept soldiering through the checklist, hoping to get at least one engine started. They were about halfway through the turn. Cockpit warning bells were chiming repetitively.

Far away in his radar room at New York Approach, the controller Patrick Harten was focusing with equal concentration. The situation was critical, with no room for error. A balance was required between offering too little assistance and distracting the crew with unnecessary talk. Even if these pilots and their passengers somehow survived, the FAA hierarchy was going into full defensive posture, and Harten's performance was going to be closely scrutinized. But that was not directly his concern at the moment. He was in top form and overdrive. He phoned an alert to LaGuardia Tower. He said, "Tower, stop your departures. We got an emergency returning."

"Who is it?"

"It's 1529. He, ah, bird strike. He lost all engines. He lost the thrust in the engines. He is returning immediately." The flight was 1549, not 1529, and Harten, like Sullenberger, was mixing the digits under stress. It did not matter.

The Tower controller was incredulous. He said, "Cactus 1529, *which* engines?"

Harten said, "He lost thrust on both engines, he said."

"Got it." Word spread immediately to multiple facilities. Traffic was stopped at LaGuardia, with departures held on the ground and arrivals vectored away from the scene. The airport crash crews were alerted, as were a host of other emergency responders, including the Port Authority and New York City Police, and soon enough the Coast Guard.

Again, there are two runways at LaGuardia. They cross. Both are fairly short, at 7,000 feet long. Between them they provide four thresholds, or directions to land. One of those thresholds is labeled Runway 4. That is the runway and direction that Sullenberger and Skiles had used for the takeoff. Its threshold is bounded by a busy expressway and embedded in a neighborhood of Queens. If you crash short of it, you and your passengers will die, and you will likely kill people on the ground. The airport's other three thresholds are built out into the upper reaches of Long Island Sound—industrial waters cluttered by bridges and causeways, and by pilings supporting the approach lights of the runways themselves. If you are going to undershoot or overshoot a runway, you'd rather not do it here.

But Harten is a controller, not a pilot, and he was not going to second-guess the crew. Sullenberger had explicitly called for a return to the airport, and Harten made the best possible offer. He radioed, "Cactus 1529, if we can get it to you, do you want to try to land Runway 13?"

Runway 13 offered the threshold closest to the airplane's position. The airplane was still in the left turn, passing through a westerly heading, with another 50 degrees to go before rolling out to the southwest. From the cockpit, LaGuardia still lay out of sight, somewhere behind and to the left, but the airplane was descending through 1,900 feet, and Sullenberger knew he was low. He answered, "We're unable. We may end up in the Hudson."

Harten did not immediately respond. Such was his mastery of his trade that he had not lost sight of his full responsibilities. He took a few seconds off to give another airline pilot a routine vector to the west. That pilot, too, knew his job. He acknowledged the clearance tightly, without reference to the emergency under way on the frequency.

During that time in Cactus 1549, Skiles was working through the checklist. He said, "Emergency electrical power. Emergency generator [is] not online."

The airplane's emergency generator is driven by the sort of drop-down windmill—the so-called ram air turbine—that had deployed during Piché's epic glide. On Flight 1549 it never came into play, for now because the left engine core was still driving a main generator and providing normal electrical power.

Sullenberger said, "It's online." He may have meant the main generator, or the auxiliary power unit in the tail. Either way, it didn't matter; they had plenty of electrical power, and the airplane's all-important control system remained fully functional. This was by no means a given in this case of a dual-engine failure in a semi-robotic airplane, and it was no small factor in the events that followed.

Part Two

FLY BY WIRE

FIVE

THE ENGINEER

Chesley Sullenberger's qualities emerged in full force during the first few seconds of the emergency over the Bronx. In retrospect, what mattered most to his ultimate success was not what he did, but what he chose not to do, his shedding of distractions, the concentration that he brought to the crisis. It was an exceptional performance, easy enough to dream up in the abstract, but extremely difficult to execute in practice. His physical control of the airplane, however, is another matter, and though nearly flawless, less reflective of unusual skill. Understanding this requires a pause in the account of the glide and a brief excursion through history. We can start with the moment when people first realized why birds have wings. The wonder now is not that our species flies, but that we waited so long to do it. Airplanes are such elegant and simple devices that in their basic form they seem less to have been invented than discovered. They require two wings and a tail, maybe some means of propulsion, and a few moveable surfaces to provide for control—nose up and down, tail side to side, and wings rolling left and right. You can sit inside them, sit outside them, and fly them upside down. If you're cold, you can heat the cabin. If you're high, you can pressurize. Basically this is what the Wright brothers figured out. By avoiding learned committees, they got the job done, and they solved the problem once and for all. Nearly every new design ever since has been just an elaboration on their thinking. This goes for Boeing 737s, supersonic F-4 Phantoms, drug-running Piper Aztecs, and little Aviat Huskys like mine. This goes for Air Force One. There are families now raising their sixth generation of pilots, each with the knowledge that all airplanes are fundamentally alike, despite variations due to speed, weight, and power, because every airplane in its soul is still a Wright Flyer.

Or almost every airplane. In the early 1970s a new strain arose, with complex roots stretching back twenty years or more. It had been obvious for quite some time that, in theory, the best way to connect cockpit controls with the moveable control surfaces on the wings and tail was not through the standard hydraulic and mechanical links (which are heavy, prone to breakage, and vulnerable to enemy fire), but rather through transducers and lightweight electrical wires. To a limited degree, and always with full mechanical backup, some such electrical control circuits had already been introduced into a few audacious designs, most notably the supersonic Concorde, which was in flight testing at the time. These were analogue machines, as airplanes had always been, and most were largely conventional in their handling. Care was taken to mimic the old mechanical systems and provide familiar feedback to the pilots—for instance, by artificially stiffening the flight controls in response to increases in airspeed. Designers believed that this was necessary to keep pilots from overstressing the airplanes. Though heated arguments on the subject continue today, with Boeing using onboard computers to mimic conventional handling in its latest unconventional designs, history has shown that the worries were unnecessary. There is, for instance, no control-stick feedback in the Airbus A320 that Sullenberger was flying, and even in conditions of degraded control, there is no history in that aircraft type of excessive loads being applied in flight.

But that's getting ahead in time. In the early 1970s the big event was the advent of lightweight digital computers, and the dawning realization that they could be wedded to electrical control circuits—to huge advantage in the conduct of flight. This marriage between electrical control circuits and digital computers was to become known as fly-by-wire. The pioneering work was done in the United States at NASA's Dryden Flight Research Center, in the deserts of California, where an old supersonic F-8 Crusader was rigged up as a test-bed, and the world's first fully digital fly-by-wire airplane took to the skies. Soon it became clear that little risk was involved. Through the use of overlapping computers and control configurations, a digital fly-by-wire system could be made so reliable that mechanical backups could be done away with entirely. The proof was mathematical: though backup electrical control systems do require electricity and are not completely immune to failure (no system is), they are theoretically more robust than any mechanical backup that can be devised. This meant that airplanes could be made lighter, safer, and cheaper to fly. But that was just the start. In execution, the new technology was revolutionary. Within the constraints of mass and momentum, it allowed for the imaginative rethinking of flight, for the invention of airplanes perfectly matched to their tasks, and for the simultaneous delivery of wish-list handling characteristics quite beyond what nature until now had allowed. It also permitted a completely new relationship between pilots and their machines, based on working at all times through the interventions of flight-control sensors and computers. For the first time in history, airplanes could be made that would be fundamentally different from the Wright brothers' Flyer.

The military saw the advantage, particularly of loosening the stability of fighters to increase their maneuverability, yet also providing pilots with airplanes that were docile to fly. The F-16 was the first such design, a radical advance over Sullenberger's old F-4, and one of the best-mannered airplanes ever built. It was called the Electric Jet at the time. Setting aside the cost to taxpayers, it was an enormous success. Other fly-by-wire designs followed, leading to the strangely shaped Stealth bombers and fighters of today, which would be completely uncontrollable were it not for the constant interventions of onboard computers. Crudely put, because of digital technology, machines have the capacity to fly (and fight) better than any human can. It is an unavoidable fact that will slowly squeeze combat pilots from their cockpits, ultimately to be left on the ground. The wisdom of this can certainly be argued. But the trend has become so evident that most of the top graduates of the U.S. Air Force Academy, looking at career advancement, are now opting out of airplanes in order to handle pilotless drones by remote control instead.

In the foreseeable future there is no risk of such a fate for airline pilots, because the public would refuse to go along. Nonetheless, many professional pilots have resisted the concept of fly-by-wire designs. Pick a date, 1985, five years after Sullenberger left the Air Force. Lockheed had just abandoned the commercial aircraft business, McDonnell Douglas was struggling to get by, and Boeing was the dominant manufacturer worldwide, with little incentive to rethink airplanes and retool its factories based on a still-immature technology. Furthermore, Boeing's constituents among airline pilots were very clear in their disdain. The pilots were having a tough enough time already. Between the turmoil caused by airline deregulation and the coming of automation, which was reducing crew sizes from three to two, the glamour of airline jobs was visibly fading, and the profession was becoming less fun, less prestigious, and less well paid with every passing year. This was the start of the trend that Sullenberger complained about to Congress after his water landing twenty-four years later. What he did not express to Congress is a paradox that has been operative all the while: even as the working conditions grew worse and salaries were cut, plenty of new people lined up for the jobs, and the safety record kept getting better. However welcome this was from a larger perspective, it gave pilots ever less leverage in their struggle against the profession's decline. Rarely is this admitted in public. Sullenberger came close during the NTSB hearing, when he said, "We make it look too easy," but he added that his reference was to complacency. When he then characterized airline flying as "never knowing when or if one might face some ultimate challenge," it was slightly embarrassing, given that test pilots were sitting in the audience. The idea that piloting is a death-defying act demanding unusual coolness and skill may be useful, but it has been obsolete for at least fifty years. By the 1980s even the public was catching on. And now what? Pilots were supposed to cede direct control of the airplane and fly via the interventions of electronic boxes programmed by smart-aleck engineers? No wonder they balked.

Boeing applauded the traditionalists' resolve. But the future came at them anyway, and from an unexpected corner—the city of Toulouse, on the Garonne River in southwestern France. Toulouse is the headquarters of Airbus, an unlikely European consortium initially cobbled together as a make-work program in the late 1960s by the governments of Britain, Germany, and France. Its mission was to compete with Boeing in the commercial airplane market. For Americans this was a joke. Airbus was seen as the very model of an ineffectual political construct, and an example of how not to go about doing business. Even within Europe it was mocked and reviled. It did not deliver its first airplane, the A300, until 1974, and it required another nine years before, in 1983, it delivered its second model, the A310, as a sawed-off version of the first. These were twin-engine cattle cars, with conventional cockpits and controls and none of the graceful lines of Boeings. The first models were toured around empty by wishful Frenchmen, carrying gourmet food and Airbus girls. The innovations of even the second-generation A310 were not enough initially to excite the market—the first twin-engine, twin-aisle, wide-bodied airliner with two-pilot crews. So what? Boeing, with all its competence, was offering the same thing at about the same time. The Airbuses sold poorly at first. Boeing chuckled in its Seattle castles. The world yawned and looked away.

But if there was a miracle over the Hudson on January 15, 2009, it relates to a larger one that occurred on the Garonne twenty-five years earlier, and has since allowed the Airbus consortium—that ineffectual socialist pipe dream—to take away nearly half of the global commercial airplane market from the United States. It was able to do this not because of unfair subsidies, as Boeing has claimed, but because of a culture of intellectual courage that existed in the 1980s within the Euro construct in Toulouse, a bet-the-farm determination to rethink airplanes from scratch and to challenge Boeing in the only way that might succeed—by leaping forward unhindered by tradition and without fear or compromise in the design. The effort was led by a charismatic French test and fighter pilot named Bernard Ziegler, now retired, who must stand as one of the great engineers of our time. Several years earlier, Ziegler had been behind the decision to go with the two-pilot cockpit in the A310, and as a result he was so despised within the French pilots' union that he received death threats and had to live under police protection for a while. It did not help that he was vocal and smart, and was himself one of the most accomplished pilots around. Now he was pushing a series of radically new airplanes that he proclaimed could be made so safe and easy to fly that they would be nearly pilot-proof. He did not say "idiot-proof," but his attitude was undiplomatic, especially in France, a country where pilots still wear their uniforms proudly. The opposition was immediate and strong, but Ziegler was not a man to back down. Fortified with his technical zeal and the determination of a few others, Airbus forged ahead with the project despite a chorus of warnings. After a period of invention, in 1983 it delivered the first A320 of thousands to come. This was the same model that Sullenberger glided toward the Hudson River a quarter century later. Without doubt, it is the most audacious civil airplane since the Wright brothers' Flyer—a narrow-bodied, twin-turbofan, medium-range jet with the approximate capacities of a Boeing 737, but with extensive use of composite materials, a brilliantly minimalist flat-screen instrument panel, sidestick controls without tactile feedback, and, at the core of the design, a no-compromise, full-on digital fly-by-wire control system that radically redefines the relationship between pilots and flight.

·

Ziegler has written a book titled *Les Cow-boys d'Airbus*, about himself and his friends. Recently I went to see him in Toulouse. He is deep into his seventies now, but still trim and athletic, and married to his original wife, a long-haired beauty in the 1970s who, he assured me, is every bit as beautiful today. Life is good. His English is thick. His French is formal without being arch. He picked me up at the airport in a hybrid electric car, which seemed appropriate to me. He had dirty burlap sacks in the back because, he said, he was about to leave for his second house, in Corsica. We drove to a country restaurant on the banks of the Garonne, where we talked about the past. He is a Gaullist of the last generation for which Gaullism makes sense, but more of a free-thinker than a conservative, as the A320 amply demonstrates. He was born into the ruling caste of French technocrats, the product of Napoleonic traditions, and engineers par excellence. His father was Henri Ziegler, a pilot and engineer educated at the elite École Polytechnique, who joined de Gaulle in London during World War II, became the commander of the Free French air force, and had many adventures during the war, including parachuting into France to coordinate with the Resistance. After the war he rebuilt Air France, went on to manage the national aerospace company, now called Aérospatiale, and in 1968 cofounded Airbus, which he served as president until his retirement in 1975. He died in 1993. He never made much money. He was a public servant. He was passionate about airplanes all his life.

Bernard was seven years old when France was occupied by German forces. With his mother and two brothers he fled Paris for a village in the Vercors, a redoubt of the Resistance in the mountains near Grenoble. As the war proceeded, the village was spared by the Germans, though others nearby were destroyed. The men had fled into the forest to join the fight. Bernard's mother was active in their support and often sent her boys into the forest carrying supplies. These memories are still alive for Ziegler. His mother was a hero, and so, at a young age, was he. He did not know that he was a hero. He thought he was just taking food into the forest. But his status became official when later his mother was awarded a Croix de Guerre, "For Resistance with Her Children." Toward the end of the war an officer in an Ameri-can uniform drove up in a jeep, sought her out, and began to kiss her. I mean, really kiss her. Ziegler was shocked until he realized that the officer was not American but French, and indeed was his father. He had not seen him for two years.

After the war the family returned to Paris. A daughter was born, Ziegler's sister, whom he describes as "a fruit of the Liberation." As a teenager, Ziegler learned to fly small airplanes in the Air France aeroclub. After graduating from a lycée in Paris, he followed his father's trail into the École Polytechnique, to study engineering. Given his family's connections, this should have placed him on a fast track toward a high management position, but he graduated so low in the ranks as assessed by the learned professors that his choices were limited, and he opted to join the French air force as a pilot. He went through flight training, became a fighter pilot, and shipped off to war in Algeria, where he flew an American-made propeller-driven T-6 in close air support, doing the usual thing of bombing and strafing rebellious peasants. One morning he was shot down. He was attacking insurgents, and flying at very low altitude, when his airplane was hit with heavy-machine-gun rounds and burst into flames. He did not have an ejection seat. He managed to zoom upward, jettison the cockpit canopy, roll inverted, and bail out. He was so low that his parachute barely had time to open before he hit the ground. This saved him, he told me, from being shot while hanging in the straps. There were fedayeen all around, and as usual they were especially angry about being attacked from the air. Once Ziegler was on the ground, they tried to get at him, but other T-6s kept them at bay until a helicopter arrived and lifted him to safety. He was shot down at 8:00 a.m., got to the hospital at 9:30 a.m., and at 10:00 a.m. hobbled on crutches across the tarmac to meet his wife, who was flying in from France for a visit. When she saw him on crutches she assumed that he had hurt himself while partying.

She said, "What did you do last night?" His answer was "Nothing. I got shot down this morning." So the question became a joke between them.

Ziegler liked to party, but he was also a serious man. Having returned to France, and while still serving as a fighter pilot, he resumed advanced engineering studies, this time at another elite school, the École Nationale Supérieur de l'Aéronautique. At the age of thirty-one, in 1964, he was assigned to the French national flight-testing center, where he worked for the next eight years, flying all manner of airplanes, including heavy airliners and the Dassault Mirage G, a prototype supersonic fighter that was his special charge. Those were interesting times. They came to an end, however, after Ziegler violated French military law in 1972 by traveling in civilian clothes and without authorization to Israel, where a friend who was attempting to manufacture a small transport airplane needed help with flight testing and design. A visiting French general spotted Ziegler in a Tel Aviv hotel and reported his presence to the command. When Ziegler returned to France, he was imprisoned for forty days and cashiered. After a military career of fifteen years, he was thirty-eight, with a family to support, and suddenly in need of a job. He was too old to start flying for Air France, and overqualified as well. Dassault considered bringing him on as an engineering pilot, but balked because of the connection to his father, the senior Ziegler, Henri, who was running both Aérospatiale and Airbus at the time. These companies were competitors to Dassault. Henri understood the bind his son was in, and he did the natural thing, offering him a position as the chief test pilot at Airbus. There would be grumblers in the ranks, but even they would have to admit that Bernard was well suited for the job.

The need for a chief test pilot arose because the first Airbus airplane, the A300, was ready to fly. Ziegler made that first flight, and many of the flights that followed, through certification and beyond. Later he did the same for the A310. As the most experienced pilot in these airplanes, he became their evangelist, and managed customer support when at last they began to sell. His father by then had long since retired, and reservations about Ziegler had been swept aside by his courage and dedication, his willingness to cut his own pay when times were tough, and the sense of mission that he nurtured within the engineering and flight-test staff. He does not brag, but others say this about him, and the record speaks for itself. He rose rapidly at Airbus, and while continuing to lead the test flying, eventually assumed overall responsibility for engineering and design. By the early 1980s his reputation was so good that he was able to walk into the office of the chief operating officer, an Airbus cofounder named Roger Béteille, and over the course of merely thirty minutes convince him to gamble the company's entire future on a risky idea: a series of audacious new airplanes that would vary in range and size, but share identical cockpits and wish-list handling characteristics based on advanced electronics and fly-by-wire control. The control system would involve redundant but dissimilar computers linked together in an intricate web of overlapping jurisdictions designed to accommodate failures and, if necessary, to defend itself from collapse by conducting staged withdrawals through the layers of its own capabilities, from the highest level of control to something resembling conventional handling. Electricity would be the requirement here, but overall, the system could be made so reliable that it would need little mechanical backup, if any at all. This was the idea that Ziegler brought to the office. Béteille saw the potential right away.

He had been through the battles over two-pilot cockpits, and must have known that any such semi-robotic airplane would encounter opposition within the airlines and unions, but he was inventive himself, and what Ziegler was proposing was an irresistibly elegant design. An electric jet? Béteille said yes. He asked only that Ziegler build the first one on budget and take it through certification in Europe and the United States within two years.

It was an ambitious request. Béteille was rushing because the global fleet of tri-engine Boeing 727s had grown obsolete, and if Airbus did not produce a correctly sized replacement soon, it would cede the business to Boeing's own offering, the well-established twin-engine 737. But how do you invent a new approach to flying, and do it overnight? Ziegler could borrow from NASA's research, and from aspects of the Concorde's design, but he was still looking at a nearly blank page. Sitting with me on the banks of the Garonne years later, he said, "Yes, and that was all the difficulty. Because the design was extremely open. Our engineers said, 'You can do what you want. What do you want?' And then we started to understand that we didn't know what exactly we wanted." But they figured it out in time, and to such an extent that the solutions they devised for the first model, Sullenberger's narrow-body, medium-range A320, were applied essentially without change to variants called the A319 and A321; to the wide-body, long-range A330; to the four-engine, ultra-long-range A340; and most recently to the jumbo two-deck A380, the world's largest airliner. With partial exception for the heavy A380, in which weight and experience have shifted some equations, all of these airplanes handle very much alike. Furthermore, they all have the same computerized cockpit, with the same little sidestick controls—unimposing joysticks, one each for the captain and copilot, that redefine the pilot's job by redefining the very nature of flight. This is a slightly complex subject, because it requires basic knowledge of conventional airplane behavior. For now it is enough to recognize that within the limits of physics and structural science, Ziegler and his colleagues identified the wrinkles of conventional handling and mostly ironed them out. The result in the A320 is the product of genius—an airplane that is highly unusual, to be sure, but exquisitely wrought, a delight to handle, and extraordinarily easy to fly.

•

I said as much to Ziegler on the banks of the Garonne. Recently I had explored the design in a full-motion simulator. He answered, "Yes, but you know sometimes I wonder if that was not our mistake. Sometimes I wonder if we made an airplane that is too easy to fly."

I thought I knew why. I said, "Why?"

"Because in a difficult airplane the crews may stay more alert." Ziegler had a sip of French wine. He said, "Sometimes I wonder if we should have built a kicker into the pilots' seats." It was an old frustration, I understood. Even in this airplane, pilots have managed to die.

Ziegler tried hard to keep it from happening. In his pursuit of perfection, he created an airplane that is not only easy to fly, but in the extreme will intervene to keep people alive. The interventions cannot be overridden or disabled. They are called flight envelope protections. They stand in the background during flight, ready to act at all times, but outside the range of normal operations. They are a fruit of fly-by-wire. Pilots can fly their entire careers without encountering them, but with their existence always in mind. In the A320, for instance, if you slam the sidestick control to the side, the airplane will roll in that direction by 15 de-

grees per second, and in two seconds will rush past the standard
30 degrees of bank, but rather than continuing the roll to stand on
a wing and then go inverted as a conventional airplane would, it
will halt the maneuver at 67 degrees—precisely the bank angle
where the airplane, if it holds altitude, will be at the maximum
2.5 G load limit of stress for its structure. No matter how disoriented you are, there is nothing you can do to make it go farther. If
at that point you pull the stick fully back, the airplane will lock
onto its present altitude, turning hard. You don't have to look outside. You can even close your eyes. If you then release the stick,
just abandon it, the airplane will roll back automatically to a more
civilized 33 degrees of bank, where it will freeze its nose attitude
to help you recover your senses.

So return now to cruising, wings-level flight. From that position, if you slam the sidestick fully back, the airplane will pitch up
rapidly, but again will impose no greater gravity load than the
maximum safe 2.5 Gs. You can be as rough as you want, and you
won't shed your wings or tail. During this maneuver, with the stick
held fully back, the airplane will not go vertical and into a loop
as any conventional airplane would, but will freeze its attitude at
30 degrees up and refuse to pitch any higher. Then, if you reverse
yourself and push the stick fully forward, the nose will pitch down
at a rate that will cause the airplane to pass through 0 G (weight-
lessness), but not exceed the negative flight load limit of –1.0 G.
Incidentally, at –1.0 G, the passengers could walk around on the
cabin ceiling, upside down in relation to the earth, but feeling
normal. This might be amusing to them if they were in the right
frame of mind. The game, however, could not be sustained for
long, because in the A320 the nose will stop dropping when it arrives at 15 degrees below the horizon. Then, even if you keep the
stick fully forward, the nose will refuse to dip any further. The

airplane is now in a dive and accelerating fast. For argument's sake, you insist on holding the stick fully forward. The airplane doesn't care. When it arrives at its high-speed limits (a Mach number at high altitude, a dynamic airspeed at low altitude), it raises its nose and, on its own initiative, pulls up safely to maintain the maximum speed. The pull comes from Ziegler's hand. You can give your passengers an exciting ride, but he is determined not to let you break his airplane.

His most elaborate protections intervene at the other extreme, when the airplane is flying slowly, mushing through the sky at high angles of attack and approaching a stall. Angles of attack are the angles at which the wings greet the air in flight. As an airplane slows, the angle of attack increases. The problem arises when suddenly the angle becomes too sharp. The airspeed at which this occurs varies depending on weight, center of gravity, G-load, and wing configuration—particularly the extension of leading-edge slats. Even with the help of those slats, at some point the air cannot make a sharp enough bend to keep flowing smoothly over the tops of the wings, and it separates away from the skin to swirl off into ineffectual eddies. That is what is meant by a stall. The term here does not refer to engines but to wings. Engines quit; wings stall. When engines quit, airplanes glide. When wings stall, airplanes fall. That's a slight exaggeration, because it depends on the airplane, with some maintaining partial lift even though the wings are stalled, and so reacting less viciously than others. In all of them, however, recovery requires that the airplane be nosed down, a maneuver that may consume hundreds of feet or more, especially in heavy swept-wing airliners. If there is not sufficient airspace below, the airplane will hit the ground with extreme violence and out of control. In such accidents, everyone dies. These are Grade A crashes. They occur typically during an approach to landing, when speeds are necessarily slow and wind shear or ice on the wings and tail may aggravate the situation.

The most recent case in U.S. airline history occurred a month after Sullenberger's glide, on the night of February 12, 2009, when a Canadian-built turboprop associated with Continental Airlines stalled while approaching Buffalo, New York. That airplane was not a fly-by-wire design, but it was equipped with an automatic stick pusher meant to protect the airplane from stalling. Significantly, the stick pusher could be overpowered by the pilots if necessary. In other words, the captain had the final authority—and on that night, he used it. When the airplane unexpectedly came close to stalling, and the stick pusher suddenly lowered the nose, the captain reacted in surprise and pulled back hard on the controls. Had he allowed the safeguard to function, he would have accepted a dip in altitude and the need to increase his airspeed during the descent to the airport. Instead, he overpowered the safeguard and caused a violent loss of control. He killed all forty-nine people aboard, himself included, and one person on the ground. Other airplanes have buzzers, clackers, horns, and stick shakers meant to warn of approaching the edge, and still, occasionally, their pilots manage to stall.

•

These are some of the pilots Ziegler had in mind when he built the A320 protections—not masters like Sullenberger and Skiles, or even average airline crews, but people at the low end of the scale, who occasionally will be at the controls of any airplane that is widely sold and flown. Unsafe pilots? Sure, of course, there are quite a few, and testing can go only so far in weeding them out. For one thing, the testers are dishonest in half of the world, and the tests themselves are incompetent. For another, the most dangerous traits in pilots may be hidden and hard to define. This is why families and friends are so often surprised when pilots crash and die. Ziegler mentioned it to me: "They say, 'But he was such a good pilot! And he had a license, you know!'" He suddenly looked fatigued; it had been such a long struggle. He said, "Can you imagine?"

I said that I could. There are more than three hundred thousand airline pilots in the world. If you design airplanes for them to fly, you must grapple not only with the existence of a few who are incompetent from the start, but also with the fact that plenty of once-excellent pilots grow unsafe with time. They become arrogant, or bored, or complacent. They drink, they fade, they erode. They become bitter about life or their bosses. Formalized cockpit teamwork does help to maintain some discipline, but because compliance is sketchy in reality, it offers poor guarantees. In the long run, personalities and national cultures seem to matter as much as experience in flight, but these are difficult traits to control for, particularly in a crowd that tends to be self-impressed and is imbued with a union mind-set that holds employment seniority to be an inviolable measure of privilege, if not of performance. Cockpits are extremely private places, and the authority of captains, though now somewhat diminished, necessarily remains strong. Cockpit voice and flight data recorders are accessed only after accidents, and they are self-erasing loops containing the final several hours at most. In normal operations, no eavesdropping is allowed, and it is difficult for airlines to know how their pilots are actually flying.

What did Ziegler want? He wanted to build an airplane that could not be stalled—not once, not ever—by any pilot at the controls. To achieve this he installed three levels of low-speed protection, based on active and ongoing calculations of the stalling angle of attack, and shown with color codes on the airspeed scales directly in front of the pilots, on the cockpit's primary flight displays. In the jargon of aviation, angle of attack is known as alpha. Ziegler's first gate therefore is known as alpha prot, with *prot* standing for "protection." The speed for it varies widely depending on weight and wing configuration, but it might typically be about 150 miles per hour, or 10 miles per hour less than a typical minimum speed for a final approach to a landing. At any altitude above treetop level, if the sidestick is standing in the standard neutral position and the airspeed strays to that point, the airplane gently noses down to keep the speed from slowing. You can override the protection by bringing the sidestick fully back. If you do, the airplane pitches up and continues to slow, now into a range where in a conventional airliner you would never dare to go. Still pitching up and slowing, you come to the next of the protections, a gate where the system assumes that the reason you are holding the stick back is that you are desperately trying to pull away from the ground, and it automatically throttles the engines to maximum thrust and retracts the speed brakes if they are extended, to put the airplane into an emergency climb. That gate is known as alpha floor. By holding the stick back, you can pass through it and continue to slow (despite the engine thrust) until you come to the final protection, known as alpha max. *Max* means "maximum": stop the speed reductions—enough. It is one place where Ziegler's system fully intervenes, balancing the airplane on the narrow edge of a stall, and keeping it there with a coolness and precision that few pilots can match. In the natural order of things, most Airbus pilots will experience such interventions only in simulators and in training. But every once in a great while, when something goes badly wrong during routine operations, the system emerges from the background to keep people safe. It is unknown how many times this has actually occurred, but certainly there are hundreds of people alive today who owe every breath they take to the arcane philosophy behind the Airbus design.

SIX

THE PROBLEM

The combination of the low-speed protections and the ability to slam the sidestick back without exceeding the 2.5 G load limit of the airplane has a profound effect on the performance of emergency pull-ups. These are the last-chance maneuvers in any airplane where pilots who unknowingly have strayed while at low altitudes in the clouds or at night are suddenly alerted to an impending crash by the ground-proximity warning system in the cockpits: *Terrain! Terrain!* and *Pull up! Pull up!* By definition, it comes as a complete surprise. Depending on the sophistication of the warning system, you may be mere seconds from impact, and in mountainous terrain your survival may depend on a sustained upward zoom by the airplane. In recent years the problem has eased because of the introduction of new systems based on geographic databases, which "look ahead" to give pilots better advance warning and more time to react. Nonetheless, emergency pull-ups will still occur, and the techniques involved offer insights into design philosophy. In the A320 the procedure is clear: you simply slam the sidestick back and hold it there. Ziegler calls this "snatching." You snatch the stick and trust in fly-by-wire. You can be quick and clumsy. You will stress but not overstress the airplane. You will not lose control. You will zoom up to the narrow edge of the maximum angle of attack, milking every last pound of lift from the wings, and without fear of a stall. If you haven't already shoved the throttles forward, the system will go to full thrust for you. The speed brakes will retract. Your control will be smooth throughout. Whether you survive is another question. But you can be sure that you have extracted maximum performance from the airplane—or that it has extracted maximum performance from itself. According to Ziegler, it hardly matters what sort of pilot you are.

In a conventional airplane it's not like that. If you respond to a ground-proximity warning by brutally snatching the controls, you will either overstress the structure or rear up steeply into a stall, and it is likely that you and your passengers will die. Because of these constraints, emergency pull-ups in such airplanes require more cautious reactions and greater piloting skills. The procedure is to raise the nose a set rate (about 3 degrees per second) to a predetermined pitch, shove the throttles forward, retract the speed brakes, and then keep raising the nose until you come to the stall warning—probably a stick shaker that rattles the controls in your hands. You then try to stay there, feeling your way in and out of the stick-shaker zone so that you can define its forward edge. This is the best you can do, the closest you dare come to a stall at low altitude. As a result of these considerations, and of the high psychological demands implicit in any emergency, it is unlikely that you will be able to exploit the airplane's full performance potential during a pull-up to get away from the ground.

The classic case in a conventional airplane occurred near Cali, Colombia, on the night of December 20, 1995, when an American Airlines Boeing 757 with 163 people aboard was arriving from Miami and descending through the darkness over mountainous terrain. The 757 is a superb interpretation of the Wright brothers' Flyer, and perhaps the most beautiful airliner ever designed. It has a long, slim fuselage, long, slim wings, and a two-crew cockpit with impressive old-fashioned control wheels, electronic flight instrumentation, and, for primary navigation, a flight management system that marries a geographic database to inertial and GPS guidance. The two pilots in the front that night were both former Air Force fighter pilots, each with more than two thousand hours of experience in this type. The captain was fifty-seven years old, nearing the end of a long and successful career. He had an excellent reputation. The copilot was thirty-nine. His reputation was equally strong. In the Air Force he had once been named Instructor of the Year. Both pilots lived in Florida. They were good family men. They were athletic. They had been schooled repeatedly in Cockpit Resource Management. If asked, they would have sworn to the need for standardization, for regulation, and for what passes as professionalism in the trade.

·

It was a black night, though some lights may have been in sight. The weather at the airport was mostly clear, with a few clouds scattered about. Cali lies in a long, narrow valley oriented north–south, about 3,200 feet above sea level, and with mountains rising to 14,000 feet on each side. The air traffic controllers there did not have radar, because the rotating dish had been destroyed in the civil war several years earlier. They had to rely, therefore, on radioed position reports from pilots in flight—a typically safe if inefficient procedure that remains the norm for air traffic control in most parts of the world. At 9:34 p.m., the American crew checked in about sixty miles north of the airport, descending southbound through 23,000 feet, with the copilot handling the principal flying duties and the airplane on autopilot. They were about two hours late because of delays in Miami. The captain illuminated the seat belt sign and asked the flight attendants to prepare the cabin for landing. He switched on the landing lights to make the airplane more visible to traffic, but the night was quiet, with no other airplanes in the vicinity.

The plan was to descend beyond the airport, turn around, and land to the north: the flight management system was set up to do that by navigating a track shown on a charted arrival procedure. The airplane was in perfect condition and position, and the situation was normal. But then the controller offered a time-saving alternative, a straight-in approach for a landing to the south on a runway labeled 19 for its compass orientation of 190 degrees. After a brief discussion, the pilots went for it. Because they were now much closer to landing, they were suddenly a little too high and fast. They throttled back sharply and deployed the speed brakes to increase the aerodynamic drag. The airplane glided down through 17,000 feet and began to slow below 350 miles per hour. All this was normal, but the situation was no longer quite under control. While the copilot monitored the autopilot, the captain rushed to pull out the chart for the expedited arrival, and he struggled to reprogram the flight management system with a sequence of waypoints called for by a new approach procedure. The waypoints were radio beacons on the ground, but overlaid by the flight management system's database, which, again, drew global guidance primarily from satellites, allowing the pilots quite appropriately to skip tuning frequencies into the airplane's now-secondary navigational radios.

At this point a chain of unfortunate events occurred. First, the flight management system obscured the fact that the airplane was farther along the new arrival track than the pilots realized, and indeed had already passed the correct entry gate for the procedure, a beacon named TULUA, after a nearby town. TULUA stands in the valley forty-three miles north of Cali's primary beacon, and a good distance north of the airport as well. The captain assumed (erroneously) that it lay ahead, but in his rush to arrive, he skipped over it. After a confused conversation with the controller, he chose another way point 2.5 miles north of the runway, and got the copilot to steer directly toward it. This was a procedural improvisation of questionable benefit or legality, but in the good weather prevailing, it was probably safe enough that night. The waypoint he chose was another beacon called ROZO. The problem was that the captain then entered the wrong identifying code for it into the flight management system, selecting an *R* as shown on the chart, when the flight management system's logic required him to select the whole name: *R-O-Z-O*. The system's logic was obscure and needlessly complicated, but it was the captain's job to know it, and he did not. The effect of the mistake was to activate navigation not to ROZO but to ROMEO, a waypoint more than 130 miles to the east, in Bogotá. The autopilot responded slavishly by making a large left turn to proceed in that direction, on a heading at right angles to the correct course down the valley to Cali. For reasons that will never be known, the pilots allowed the turn to happen, despite the fact that it obviously made no sense at all. Everything about that flight was south: south from Miami, south across Cuba and the Caribbean, south into South America, south during the initial descent, south down the Cali Valley on a straight-in approach to a south landing. South south south. But they let the airplane turn east. Worse, they kept descending. If you back up to a moment just before the turn, it sounded like this:

The copilot said, "Okay, so we're cleared down to five [thousand] now?"

The captain answered, "That's right, and . . . off ROZO . . . which I'll tune in here." He entered the *R* into the flight management system. He said, ". . . see what I get . . ."

The copilot said, "Yeah."

What they got was the turn to the east, a course deviation that took them sailing over unseen mountains and down into a high valley parallel to the deeper valley where Cali lies. All was blackness outside. On the ground, the controller was unaware of the turn, because he lacked the radar to see it. In the air, the pilots were descending with determination. The captain checked the chart. He said, "Twenty-one miles at five thousand's part of the approach, okay?" It was part of the approach if you were actually on the approach, but, again, the approach goes south, and they were going east.

The copilot said, "Okay."

The captain set a navigational radio for backup information. He said, "Off ULQ . . . So let me put ULQ in here, [frequency] seventeen-seven, 'cause I want to be on raw data with you." ULQ was the identifier for TULUA, the gateway they had passed by without realizing at the start of their misadventure. Belatedly, the captain was doing what he should have done before allowing the turn to the east. He was uncomfortable with the functioning of the flight management system and was reverting to basic radio navigation, setting up an independent information source at a time during the arrival when, as U.S. investigators later wrote, the flight management system's automation "became confusing and demanded an excessive workload in a critical phase of the flight."

The controller radioed, "American 965, distance now?"

The captain answered, "Uhh, what did you want, sir?"

"Distance. DME."

The captain radioed, "Okay, the distance from, uh, Cali, is, uh, thirty-eight."

The controller radioed, "Roger."

The copilot became openly uncertain. He said, "Uh, where are we? We're going out to . . ."

The captain did not know. He gave up on ROZO and decided to proceed to the entry gate. He said, "Let's go right to, uh, TU-LUA first of all, okay?"

"Yeah, where're we headed?"

Sullenberger demonstrated in his own crisis years later that he would not have messed up like this. But certainly this crew believed they would not have, either. They were experienced, current, and licensed. Their fellow pilots trusted in them. They had flawless records. In routine operations they were smooth. They knew the procedures and regulations. Throughout their careers they had always performed well in tests. They could fly by the book if they had to. But once confusion begins to dominate a crew, a chain reaction may occur, with confusion begetting confusion and blossoming beyond control. Forward motion is the essence of flight. Once you get behind in a cockpit, it is inherently difficult to catch up. You cannot hit "Pause." You cannot just pull over and stop.

The captain said, "Seventeen-seven, ULQ, uh, I don't know what's this ULQ?" In retrospect it was a disturbing question. ULQ was the identifier for TULUA, which he had just tuned in. It was printed clearly on the chart. But the navigational instruments showed it behind and to the left, when he believed it should have been ahead and to the right. He said, "What the? What happened here?"

The copilot said, "Manual . . ." Apparently he meant, Let's get off the elaborations.

The captain said, "Let's come to the right a little bit."

Referring to the controller, the copilot said, "He's wanting to know where we're headed." The copilot rolled the airplane into a tentative right turn, descending through 13,600 feet. They were dropping below the level of the highest peaks and veering in the darkness.

The captain had not yet begun to swear. He said, "ULQ . . . I'm gonna give you direct TULUA."

"Okay."

The captain punched up a navigational display for the copilot. He said, "Right, now. Okay, you got it?"

"Okay."

"And . . . It's on your map . . . Should be . . ."

The copilot said, "Yeah, [but] it's a left, uh, left turn."

The captain resisted. He said, "Yeah, I gotta identify that bastard, though . . ." Why was TULUA showing up to the left? He listened to the beeping of the station's Morse code, which came in clearly over the speakers. Dit-dit-dah, dit-dah-dit-dit, dah-dah-dit-dah. ULQ. You didn't need to be an expert, because the dots and dashes were printed clearly on the chart. He said, "Okay, I'm getting it. [Frequency] seventeen-seven. Just doesn't look right on mine. I don't know why."

The copilot was more prone to believe the instruments. He said, "Left turn. So you want a left turn back around to ULQ." He was not asking. He was telling. He rolled the airplane into a left turn.

The captain said, "Nawww. Hell no, let's press on to . . ."

The copilot was at a loss. He said, "Well, we're . . . Press on to where, though?" The airplane came out of the left turn and rolled again to the right.

The captain said, "TULUA."

But, again, TULUA was to the left, not the right.

The copilot gave in. He said, "That's a right U . . . U . . ."

It would have been a right U plus some, indeed nearly three quarters of a circle, the long way around to turn to get eventually to TULUA. They never even tried. Right? Left? A course reversal? Their confusion was complete. And still they kept descending, losing 2,700 feet per minute. It will never be known why they insisted. Had they climbed above 14,000 feet they would have had all the time in the world to figure things out and get back on track. Colombian officials later suggested that they might have been unaware of the region's topography—the copilot because this was his first trip to Cali, and the captain because he had previously flown there only at night and perhaps had never actually seen the mountains. Terrain was indeed not shown on the cockpit screens or the approach chart. But such ignorance seems implausible nonetheless. The captain had made thirteen trips to Cali—and he did not realize that it sits in the Andes? So no, they knew about the mountains. But the captain was presumptuous, if not arrogant, and both men were collapsing under the psychological stress. They must have been aghast at suddenly finding themselves in this predicament, lost in the night during an approach to landing, carrying 161 trusting souls in the back. These were top-tier pilots—or so they had reason to believe—and a climb at this point would have been an admission of their failure.

For whatever reason, they did not throttle up. Instead, the captain said, "Where're we going? One two. Come to the right. Let's go to Cali. First of all, let's . . . We got screwed up here, didn't we?"

"Yeah."

"Go direct." They were still relying on the flight management system. The captain entered the identifier for the Cali beacon. "C . . . L . . . O . . ." He said, "How did we get screwed up here? Come to the right, right now. Come to the right. Right now."

Intelligence is not a prerequisite for safe flying, but an acceptance of human fallibility is, and the two are generally linked. Ziegler mentioned it on the banks of the Garonne. He has seen such variations over the years. He said that the mark of the great pilots he has known is that they admitted in advance to their capacity for error, and they addressed their mistakes vigorously after making them. He said, *"Vous savez, monsieur. L'erreur est humaine."* Actually the Latin original, in full, goes, *Errare humanum est, sed perseverare diabolicum.* To err is human, but to persist is diabolical. Maybe it should be posted in polling stations. Certainly it should be posted in cockpits. The captain was having a hard time with it that night. He never admitted that he had screwed up. He never even admitted that he and the copilot together had screwed up. Instead he said that they had gotten screwed up, as if it had been done to them by outside forces—presumably some mysterious equipment failure. The distinction may seem like a semantic quibble, but it fits into larger patterns at play that night and helps to explain the ongoing and maddening descent. Even now the captain did not fully accept what the navigational instruments showed—that they had overshot the entry gate, that they had proceeded into uncharted territory far to the east of the final approach course, and that after all these years spent flying airplanes, this time his mental map was wrong. He was intellectually arrogant. It was diabolically stupid of him. He kept thinking they could salvage the arrival.

The copilot answered his request to "come right" by dialing the turn into the autopilot. He said, "Yeah, we're, we're in a heading-select to the right." The airplane rolled into a right

20-degree bank, with its descent rate tapering slightly and its airspeed slowing below 280 miles per hour.

The captain got on the radio to the controller. "And American, uh, thirty-eight miles north of Cali, and you want us to go to TULUA, and then do the ROZO, uh, to uh, the runway, right? To Runway 19?" It was an unhappy transmission. They had hashed this out before. Furthermore, the airplane was not north of Cali, as he reported, but closer to northeast. He was continuing to discount the displays, which showed that Cali now lay well to the right side of south.

The controller was growing uneasy. Weren't these pilots already deep into the approach? He did not speak sufficient English to express his doubts. He stuck instead to the technicalities. "American 965, you can land [on] Runway 19. You can use Runway 19. What is your altitude and DME [distance] from Cali?"

"Okay, we're thirty-seven [miles] DME at ten thousand feet."

They were thirty-seven miles north? That put them south of TULUA, and toward the airport. Why was the pilot still bringing up TULUA? Also, he had first reported being thirty-eight miles out more than two minutes before, and now he was reporting at thirty-seven miles. Presumably the airplane was traveling at least four miles per minute, perhaps five. If the distance reports were accurate, the crew had either flown in an unauthorized circle or had set off an improvised arc at an initial right angle to the straight-in approach. But if they had gone off on an arc, they wouldn't be to the north anymore. And why would they do that anyway? What was happening out there in the night? The controller did not know how to ask. Instead he answered stiffly, "Roger. Report, uh, five thousand [feet], and, uh, final to Runway 19."

Five thousand feet at twenty-one miles, and slowed—that

was the goal that the crew had in mind. They kept at their determined descent, with the throttles back and the speed brakes fully
deployed. The captain was having his own troubles with English
now. He did not acknowledge the controller's request. To the
copilot he said, "You're okay. You're in good shape now. We're
headin' . . . Shit . . . You wanna take the [Runway] 19 yet? Come
to the right, come come right to Cali for now, okay?"

"Okay."

No, it was not okay. "Come come to Cali" meant that the
captain was at last aware that Cali lay to the southwest—but he
remained stubbornly disoriented about their own position and
continued to fuss with the TULUA beacon, which, despite all the
evidence to the contrary, he believed still lay ahead. He said, "It's
that fucking TULUA I'm not getting for some reason . . . See, I
can't get . . . okay now . . . no, TULUA's fucked up."

Try going for a jog at night, closing your fucking eyes, then
faulting your fucking flashlight for failing. That's what it was like.
The copilot may have been annoyed. He said, "Okay, yeah."

The captain said, "But I can put it in the box if you want it."

The copilot refused to participate. Call it Cockpit Resource
Management. He said, "I don't want TULUA. Let's just go to the
extended centerline of, uh . . ." He wanted to angle still more to
the right to fly a straight-in visual approach to Runway 19.

The captain got it. He said, "Which is ROZO."

The copilot said, "ROZO."

"Why don't you just go direct to ROZO then, all right?"

"Okay, yeah."

This was the same waypoint that the captain had entered in-
correctly into the flight management system, provoking the excursion to the east. But there was an alternative way to define it
now, with an old-fashioned direction finder that could be used to

home in on a radio beacon there. The captain may have selected the frequency. Speaking of the display, he said, "I'm going to put that over [to] you."

They were passing into the lower altitudes, where barometric calibration of their flight levels mattered more. The copilot said, "Get some altimeters, we're out of uh, ten now."

The captain said, "All right."

The controller had similar thoughts in mind. He radioed, "American 965, altitude?"

The captain answered, "965, nine thousand feet." The airplane began to roll out of the right turn.

The controller said, "Roger, distance now?"

The answer would have been thirty-three miles, but just at that time the cockpit's ground-proximity warning erupted. It said, *Terrain! Terrain!* and made a whooping sound. The airplane was heading southwest at 270 miles per hour, with its wings nearly level, descending at 1,500 feet a minute. *Whoop! Whoop!* The ground at that moment was 1,470 feet below, as measured by the airplane's radar altimeter, but it was rising rapidly into the dark mass of a mountain slope ahead.

The mountain is called El Deluvio. It is about 9,000 feet high.

Miami lay three and a half hours behind. Presumably the airplane's forward beams illuminated the slope as the cockpit surged toward it. The beams were like flashlights. If they played on the slopes, they mottled the colors. The captain said, "Oh shit!" and shoved the throttles forward even as the copilot disengaged the autopilot and hauled his control wheel back with both hands, pitching the Boeing up and pulling Gs. The long, slim wings arched up but did not fold. Strapped into his beloved seat, the captain was full of exhortations. He said, "Pull up, baby!" whether to the airplane or his fellow pilot. There was no time for intervention. The copilot was going to fly that thing. He hauled the nose 31 degrees up and got to the stick shaker. He said, "It's okay." The warning system said, *Pull Up!* The captain said, "Okay, easy does it, easy does it." The copilot eased the control wheel forward, slightly lowering the nose. Their distance from the slope was 500, then 150 feet. They were not going to make it. The copilot said, "Nope." God knows what the passengers were thinking. The stick shaker stopped. The captain said, "Up, baby!" The stick shaker started again and did not stop. The slope was very close. The captain said, "More! More!" The copilot said, "Okay." The captain said, "Up, up, up!" The warning sounded, *Whoop! Whoop! Pull up!*

The Boeing hit the mountain nose high and climbing, merely two hundred feet from the top. It slammed through the trees on a trajectory that carried it over the top to a resting place on the far side, where a small fire broke out. For the controller the impact sounded like radio silence. He made several calls and gave up after a while when he got no response. At dawn the wreckage was found by a helicopter crew. A dog had survived, as had five severely injured passengers, one of whom later died in a hospital. As the investigation began, it became apparent that the speed brakes had remained fully extended throughout. Retraction of the speed brakes in a 757 is not without complications: in combination with the application of full thrust, it may cause pilots to overcontrol the pitch and go through large oscillations as they try to define the forward edge of the stick-shaker zone and the maximum angle for an emergency climb. For whatever reason, American Airlines did not include retraction of the speed brakes in its training for the maneuver. Nonetheless, it is obvious that leaving the speed brakes extended reduces any airplane's climb performance, and investigators concluded in this case that the crew had simply neglected to retract them. It's hard to fault them for the omission at that moment of extreme stress. But studies later indicated that if they had retracted the speed brakes at the start of the pull-up, they might have been able, just barely, to stay out of the trees. Furthermore, had they also held the optimum climb angle without fishing for it as they did (up-down-up-up-up), the airplane would have cleared the mountaintop by a hundred feet. No technology can protect passengers from such pilots, but in this particular case, had they been in a fly-by-wire design, it seems likely that everyone would have survived.

·

These were the sorts of operational realities that Ziegler confronted when inventing the Airbus protections. Experiments later found that perhaps 10 percent of airline pilots can (on a good day) squeeze the maximum performance from conventional airplanes during emergency maneuvers like pull-ups that require them to go to the very edge of flight. Ziegler was not building protections for them, but for all the others, 90 percent of airline pilots worldwide. He could not keep crews from descending on autopilot into the Andes at night, nor could he keep such crews from crashing, but once the autopilot is off and the pilots' hands are on the controls, he could guarantee the same performance to everyone—from the top 10 percent to the Cali crew, and all the pilots in between.

One might think that the concept of flight envelope protections would be welcome, but many pilots find them hard to accept. To the press, in aviation circles, and on the Internet, critics call the fly-by-wire Airbuses dangerous airplanes because they cede so much control to computers. The critics include the protections as part of the problem. They cannot argue against the stall protection, since it allows them to go closer to the edge than ever before, nor can they argue against the high-speed protection, since there is no conceivable justification in an airliner for wanting to exceed the design speed limits. But they object to the bank-angle protections, because any airplane can in theory be flown through any bank angle, and indeed can perform full rolls. They object to the pitch protections, because any airplane can in theory be flown through any nose attitude up or down, and indeed can perform loops. And they object to the G-load protections, because the 2.5-G barrier is structural certification standard and not a sharp limit, as has occasionally been shown by airliners that have pulled more than 3.5 Gs without shedding parts. Ultimately, however, what they really object to are not the protections themselves, but the fact that in the Airbus they cannot be overridden.

The Airbus is indeed an uncompromising design. Boeing has taken a softer approach with its own fly-by-wire airplanes, the 777 and 787, which artificially re-create conventional handling characteristics and allow flight envelope protections to be overridden. Ziegler says that Boeing is pandering to pilots' prejudices, and purely for commercial reasons: he cannot imagine any other reason why. He believes that if you allow crews to override the protections, you might as well provide no protections at all, because the very circumstances that cause pilots to bump up against the limits will cause those same pilots to go right on through the limits if given the chance. This was the core of Ziegler's design philosophy, and when he came under attack—well, he was used to that from the two-pilot cockpit fight, and he was not about to back down.

The quarrel mellowed with time as the fleet of fly-by-wire Airbuses grew, but even today, twenty-five years after the introduction of the A320, it continues to arise in the aftermath of any Airbus accident. Speaking of his critics among airline pilots, Ziegler said to me, "If you want to fly as they say they do, then go fly gliders, become test pilots, for all I care go to the moon. But flying for the airlines is not supposed to be an adventure. From takeoff to landing, the autopilots handle the controls. This is routine. In a Boeing as much as an Airbus. And they make better work of it than any pilot can. You're not supposed to be the blue-eyed hero here. Your job is to make decisions, to stay awake, and to know which buttons to push and when. Your job is to manage the systems."

I said, "And to take the controls in an emergency."

"In some emergencies, yes. Landing in the Hudson River, okay. But after you take the controls we give you sixty-seven degrees of bank, thirty degrees of nose up, fifteen degrees of nose down, high-speed protections, stall protections, an airplane that will not go into a spiral dive, automatic rudder coordination, automatic trim, automatic thrust, and an extremely stable platform to fly. We give you guarantees so you can react as fast as you want without having to worry about breaking the airplane. You say you want to pull more than 2.5 Gs? In fighters I have pulled 9 Gs myself, sure, but 2.5 Gs is already a lot for a transport airplane. Among all the airline pilots we flew with over the years—on familiarization flights, on test flights—never once did we meet one who was willing to pull 2.5 Gs. Never! Even though we asked them to and were sitting beside them in the cockpits: in the end we always had to do it for them, to teach them faith in the protections."

I mentioned that they might react differently if they were about to hit the ground. He answered that Airbus had done a study of high-G incidents going back for years, and had found only one case of a pilot overstressing a conventional airplane during an attempt to pull away from terrain. It occurred in July 1979, in Rio de Janeiro, Brazil, when a crew departing on a ferry flight to Senegal in an empty Lufthansa Boeing 707 received a ground-proximity warning during an excessively fast and flat climb from Rio's runway. Part of the fault lay with a controller who had been distracted by other traffic after pointing the flight toward high terrain and restricting its altitude. During their attempt to survive after receiving the proximity warning, the pilots put the airplane into an emergency turn and pulled so hard (and from an initially high airspeed) that the load built to 3.6 Gs before the airplane put an end to the story by slamming into a mountainside. When I pointed out the obvious—that the tough old Boeing had nonetheless hung together at a higher load factor than the maximum 2.5 Gs that computers allow the Airbus—Ziegler had a ready response. He said that the crew had felt their way gingerly to the 3.6 Gs, requiring sixteen seconds to get there, whereas with his fly-by-wire guarantees, they could have snatched the stick and begun a decisive climb much earlier. Be that as it may, in every other case that Airbus found of a conventional airliner exceeding 2.5 Gs, it was not in response to ground-proximity warnings, but during high-speed recovery attempts after upsets in flight—one form or another of a high-altitude loss of control. Sometimes those high-G recoveries had broken the airplanes in flight, and other times they had not. But either way, with flight envelope protections the airplanes would not have gone out of control in the first place.

At least in theory.

But all systems can fail, no matter how cleverly they are designed. Reliability is measured in probabilities, not absolutes, and Ziegler knows it well. What Airbus calculated, and demonstrated to the responsible agencies in Europe and the United States, is that the chance of a complete failure of the fly-by-wire system lies in the one-in-three-billion range. On the basis of that infinitesimal chance, the A320 is equipped with the merest excuse for a mechanical flight-control backup, and the latest design, of the giant A380, includes no mechanical backup of any kind. The confidence would so far appear to be justified: despite many assertions to the contrary, there is no known case of a complete control-system failure in an Airbus. On one occasion the captain of an A320 inbound to Kosovo reacted to a glitch by pulling enough circuit breakers to shut the entire system down, leaving the airplane dangling by a thin thread of control until the copilot was able to reset the computers. But that can hardly be held against the design. As for partial failures—which have indeed occurred—they have been accommodated by the system's multiple redundancies. Engineers, however, are not omniscient, and there is always the chance of the unforeseen. On October 7, 2008, for instance, an Australian A330 that was cruising in good weather suffered a computer failure, and reacted by pitching down so violently (at nearly -1 G) that twelve occupants were seriously injured. A computer failure should not have had that effect, but it did. The captain regained control after losing 650 feet, experienced another pitch-down five minutes later, regained control again, diverted to an airport, and made a safe emergency landing.

More recent is the mysterious case of Air France Flight 447, which plummeted into the Atlantic on the night of June 1, 2009, when flying from Brazil to France. It was an Airbus A330 with 288 people aboard, all of whom died. Because it crashed into deep waters and its recorders have not been found, very little is known about what happened. There was no bomb. There was no missile. Whatever went wrong, the disaster took a while to unfold, but the pilots remained silent on the radio. The airplane was cruising at 35,000 feet across a band of heavy but routine equatorial storms. The ride was rough. There was lightning around, but this does not seem to have been a factor. Over a five-minute period the air-

plane experienced a string of little failures, news of which it automatically transmitted to a computer in France as maintenance
items. The messages were not necessarily sent in the order that
the failures occurred—a complicating factor in trying to recon-
struct a plausible scenario. What is known is that the airplane lost
its indications of airspeed, probably because of ice on the external
probes. In programmed response, the control system retreated
one level of performance, and no longer provided fly-by-wire protections. But so what? It is almost inconceivable that Air France
pilots would have lost control because of any of this. But it is
equally difficult to relate the few known facts to the specifics of a
fly-by-wire system picking just this moment after a long fleet history, and in the midst of powerful storms, to go berserk and take
an airplane down. It appears that the airplane did not break up in
flight. It appears that the cabin depressurized. There is plenty of
speculation. Eventually the answers may be known.

Whatever happened was extremely rare. Was this the first
catastrophic collapse of a civilian fly-by-wire system? It may prove
to have been. But during the period since the design's introduction, far more passengers have been killed because of pilots than
because of airplane failures of any kind. As an airplane creator,
you have to make choices. That is why Ziegler has stood his ground
for twenty-five years, not on a self-righteous crusade to save lives,
but in a struggle for frankness where frankness is rare. To me he
said, "Bah, these airline pilots. It's all about their command authority in their minds. They say they want to be able to override
the protections? What that means in practice is they want to be
able to choose how their passengers die." He was not as hostile at
it sounds. He was making a point as if I did not already know. He
said, "First, you have to understand the mentality."

"Do you really think they are so arrogant?"

"Some, yes. And they have the flaw of being too well paid."

"So there must be no problem anymore in the United States."

But Ziegler was serious. He said, "Second, the unions' position is that pilots are always perfect. Working pilots are perfect, and dead pilots are, too."

If pilots are perfect, the problems lie with designs. Ziegler said that in 1953, when an Air France crew flew a perfectly good Constellation into a mountain during a routine descent to the airport at Nice, his father, as the airline's managing director, went with the chief pilot to report to the French prime minister. The prime minister opened by saying, "What did your pilot do wrong?" and the chief pilot answered, "Sir, the pilot is never wrong."

So it was the mountain's fault. Ziegler smiled. He is impatient with lesser minds. The problem is not that airline pilots are overpaid, or that they believe their colleagues are flawless (far from it), but that their fluency in flight combines with the routine nature of their jobs to inhibit their powers of self-assessment. Ultimately this is what lies at the root of most accidents. These are not test pilots. They are not paid to be blue-eyed heroes, to go to the moon, or to fly gliders. The tests they take are graded passfail, with ample opportunities for remedial training if they happen to slip up. Once pilots wash into an airline job, very few wash out. Assuming that the airline survives (and discounting the boomand-bust employment swings), they rise through the ranks purely on the basis of seniority. Though they do occasionally train for upsets and emergencies, their expected responses are rigidly procedural. The uniformity of even their cabin announcements gives a taste of the scripting imposed on them from above. They are not supposed to improvise. More fundamentally, they spend year after year deep inside the flight envelope, within the narrow range of maneuver that delivers smooth and safe rides to the passen-

gers. They are good at that job—and often superb. But the consequence of working so far from the extremes is to allow almost all
of them to believe that they are full masters of flying, when only a
small percentage of them actually are. By implying the truth,
Ziegler's design offended many pilots' pride. More important, by
identifying pilots as the weak link in flight, and overriding them if
necessary to protect itself, the system challenged the core public
construct of the profession and deeply threatened the unions,
whose power was already being undermined.

SEVEN

THE PARADOX

With the wisdom that experience has provided, Ziegler knows that in some way he was naïve. There was a time during the development of the A320 when he believed that with all the assistance the control system would provide, the new airplane would be nearly uncrashable. He never quite claimed it in public, but there was certainly an element of hubris in his promotion of the design. He admitted this to me in conversation, after I mentioned that his cherished airplanes—though safe—have not proved to be any safer than conventional Boeings. One reason is that Boeings, too, are well engineered and easy to handle. Another is that they are rarely flown with Cali-style incompetence to the extremes where the advantages of flight envelope protections might come into play. But there is also a negative element, a paradox that pertains particularly to the Airbus and its fly-by-wire design. It is the fundamental twist in human nature that causes people to take increased risks in direct reaction to feeling especially safe. Call it the Titanic Effect. If you believe that your ship is practically unsinkable, you might start charging across oceans of icebergs—and later wish that you had not. It has been proved time and again in history, in the conduct of public and private life, and in aviation, too. The danger of claiming that an airplane is unusually safe has always been that pilots will then go out of their way to prove you wrong.

Ziegler should have held his tongue. The A320 entered airline service in March 1988, and merely three months later the first one crashed. It was Sunday, June 26, 1988. The airplane was the fifth off the production line, the third A320 acquired by Air France. It had entered service merely three days before and had been chartered for a fund-raising "aerial baptism" flight by an aeroclub near Mulhouse, a dreary French city near the Swiss and German borders, on the upper Rhine. The aeroclub was holding an air show there at a small secondary airport with a short grass runway in a beech forest outside a town called Habsheim. The plan was to fly there from Mulhouse's commercial airport, a few miles to the south, and make two low passes for the air show crowd before heading off to the Alps to circle Mont Blanc. The first pass was to be slow, the second fast. There were 136 people aboard. At the controls were two senior Air France pilots, both with unblemished records, and between them more than 21,000 hours of flying time. The captain was an Airbus convert named Michel Asseline, age forty-four, who after years in command of Boeings had embraced this revolutionary design, and was so enthusiastic about it that he had been charged with setting up the airline's A320 program and integrating the airplanes into the larger Air France fleet. He was thought to be an excellent pilot, if a bit cocky. Some people called him Rambo behind his back. But Ziegler knew him as an important ally in the struggle for the airplane's acceptance.

It was a hazy afternoon in Mulhouse, with sunshine breaking through the clouds. As the airplane taxied for takeoff at the city's commercial airport, the mood in the cockpit was festive. Asseline bantered with a woman friend, a flight attendant who worked for another airline and was seated in the cockpit jump seat for the ride. With the copilot he reviewed his plan for the first low pass at the air show: he was going to rely on the flight envelope protections to show the crowd just how slowly the airplane could fly. To pull off the stunt, he would extend the landing gear, select full flaps and slats, throttle the engines back to idle, descend to one hundred feet above the grass runway, and then hold the altitude by allowing the nose to rotate up and the airplane to decelerate, until coming to alpha prot, the first of Ziegler's high-angle-of-attack gates, where, with the sidestick in the neutral position, the airplane would normally lower its own nose, to keep from getting slower. Asseline was not going to give Ziegler the chance. Instead he would continue slowing by holding the sidestick back. Normally the airplane within a few seconds would come to the next of Ziegler's gates, the alpha floor, where the engines would automatically throttle up to full power, but Asseline had an answer for that, too. He intended to disable the auto-throttle function by pushing a certain button, in order to pass unhindered to the absolute limit, alpha max, where by continuing to hold the sidestick back, he could briefly hang the airplane at an extreme angle of attack, balancing before the spectators on the thin edge of a stall. He would then shove the throttles forward and climb away from the ground. The copilot may have looked a little dubious. Asseline assured him (and the admiring woman in the cockpit) that he had hung the A320 at alpha max twenty times before. He did not seem to consider it important that he had done this at high altitude during airline-style training, with ample margins for recovery, in empty airplanes, and with the engines spooled up to significant power in order to maintain altitude indefinitely in the limitless airspace ahead. Apparently he was not concerned that he was going to perform the trick this time at idle thrust, at the bottom of a descent, just barely above the ground, and with a nearly full load of innocent people coming along for the ride.

Because the flight had been offered as an aerial baptism, most of the passengers were inexperienced fliers, and a greater-than-usual number had never flown before. Asseline did not ask them if they were willing to participate in his air show demonstration. If he had, few would have known to say no. He taxied the airplane onto the runway, took off, turned to the north, and soon leveled a thousand feet above a four-lane autoroute, which he knew would lead to the small airport where the air show crowd awaited his arrival. The air was smooth. The passengers remained strapped into their seats. Soon the pilots spotted the grass runway ahead. It was a short slash in the forest, with electrical power pylons standing some distance on the far side. Of course Asseline was doing the flying. He extended the landing gear, selected full flaps, brought the throttles back, and put the airplane into a decisive descent while slowing. As planned, he suppressed the protection that would normally throttle up the engines at alpha floor.

The ground-proximity warning system sounded. *Too low! Terrain!*

The copilot said, "Okay!" and switched the system off.

Ziegler at this moment was in Toulouse. Sullenberger was somewhere in the United States, fourteen years before he would first fly by wire. The airplane's radar altimeter announced the altitude in feet above the ground. It said, *Two hundred*. The copilot commented about an Air France colleague whose job was to impose standards for safe flying. Later this was interpreted as an oblique warning. As such, it was lost on Asseline.

As they approached the target altitude, the copilot said, "Good, you're arriving at a hundred feet . . ."

The altimeter said, *One hundred*.

But Asseline was keeping his own counsel, and he continued to descend—at a cocky 600 feet per minute. This quickly did the trick. He pulled out of the descent with the altimeter announcing, *Forty . . . Fifty . . . Forty*. The airplane was now just above the grass runway, flying very slowly and entering the range of Ziegler's stall protections. Asseline held the sidestick back, convinced that the airplane would not crash. Thinking out loud, he said, "Good, I definitely disengaged the auto-throttle . . ." The reference was not to the alpha floor protection but to choices in available modes of setting engine thrust in normal flight. He wanted manual control of the throttles, for going to full power when it would be required, a bit farther down the runway. He was exhibiting some confusion here about the airplane's design, since advancing the throttles fully forward will elicit maximum thrust from the engines no matter what the selected mode. The confusion, however, was irrelevant—or of no consequence to the event that followed.

The copilot was uncomfortable with Asseline's flying for other reasons. The airplane was just too damned low. He said, "Be careful of the pylons ahead. You see them?"

Asseline said, "Yes, yes . . . Don't worry."

It is not known if his female friend was impressed. But if Asseline's purpose was to impress the people of Mulhouse, he certainly succeeded. There were fifteen thousand spectators in the crowd. The marvelous new airplane passed directly in front of them, flying so slowly and with its nose so high that it seemed capable of clawing at the air to stay in the sky. The speed decreased to 128 miles per hour. Inside the cockpit, the nose-high attitude limited the forward view, and may have obscured the fact that the airplane was closing in on the trees at the runway's far end. There may have been other factors. The short grass runway was only

2,000 feet long and far shorter than pilots are accustomed to see-ing from the cockpits of airliners at that height above the ground. Furthermore, the airport had a classically shaped but miniature control tower, the sight of which may have distorted Asseline's per-ception of scale. More likely, though, he knew his position pre-cisely and was just a bit too determined to show the right stuff.

The timing was tight. One second after telling the copilot not to worry, Asseline slapped the throttles forward to the full-thrust setting. But jet engines take a while to spin up. In the A320 they require five seconds to accelerate from idle to 80 percent thrust, and an additional three seconds to get to maximum power. Eight seconds from idle to full thrust. One thousand one . . . two . . . three . . . four . . . five. One thousand six . . . seven . . . eight. De-lays vary slightly between designs, but they are normal for all jet engines, and must be taken into account. Asseline knew it full well. But having intentionally maneuvered the Airbus into a criti-cal corner, with no spare energy in the wings and no margins for error, he got the timing wrong.

He advanced the throttles at 34 seconds past 2:48 on a festive summer afternoon. Afterward there was nothing to do but wait. He seemed to know he was in trouble. He pulled the sidestick against the stops, hoping to zoom upward, trading speed for alti-tude, and buying time for the engines to accelerate. The problem was the wings had no speed left to give—and the system knew it perfectly well. Had the airplane responded to Asseline's inputs by raising its nose, it would have stalled and crashed out of control, likely killing everyone aboard. The flight envelope protections would not allow that to happen. Paradoxically, it was because of the protections that Asseline had gotten himself into this predica-ment in the first place. He would never have dared in a Boeing. This was the twist that Ziegler had not considered. In any case, the Airbus did not raise its nose in response to Asseline's demand. Instead it gently settled lower, completely under control, but with no other gifts left to offer. Fly-by-wire is not magic. It cannot defy the physics of mass and motion. The altimeter said, *Thirty . . . Thirty . . .*

The copilot implored the engines. He said, "TOGA!," the term for full power. He was praying for rain. Sometimes that works. But they were only 2.5 seconds into the 8.0-second delay, and practically nothing was happening yet.

The altimeter said, *Thirty*.

The copilot continued with his supplications. He said, "Go-around track!"

They were nearly five seconds in, and the engines were accelerating audibly. But just then the airplane's rear end began to drag through the forest's treetops. It was a lesson in life as much as in physics. This is how you lose unlosable wars. This is how you sink the *Titanic*. In the cockpit the pilots could hear the impacts. Asseline had time to say, "Shit!" The engines were powering up through 82 percent thrust, but the forest had embraced them and was pulling them down.

The embrace was gentle at first, but just a little too strong for the engines to overcome. For the air show spectators, the sight was surreal. First the airplane sailed by them almost within reach, with some announcer finding things to say. Then they watched it sail away and, without the slightest urgency, continue smoothly into the trees. Lifted by its wings, and still largely under control, it sank slowly from sight with its nose held high, until only the nose was visible moving forward through the forest like the head of a swimmer refusing to drown. In a video recording a camera tracks the low pass, captures the sound of the engines spooling up, records the heroic swim, lingers on the forest where the Airbus disappears, and, three seconds later, frames a fireball blossoming into the sky. A woman says, "*Oh la!*" Another says, "*Oh yah!*" A man says, "*Oh non! Oh non! Oh non! Oh non!*" And the announcer, like Asseline, says, "Shit."

It happened in a dream. Inside the cabin the passengers had no time to brace. After clipping through the highest branches with little damage to the tail, the airplane drifted lower into the forest and took heavier hits as it began to break and topple trees. During that period the engines briefly achieved 91 percent thrust and blasted forest debris backward, opposite to the direction of travel. Ziegler's beloved Airbus slowed violently, its wings no longer lifting, its engines succumbing. The passengers jackknifed across their lap belts and slammed their heads against the seat backs in front of them. They were being injured and bloodied, but none with sufficient force to lose consciousness or die. Left, right, and slamming again, you would have had to film them in slow motion to track the movements. The airplane was shedding sections of the tail and wings. Up in the cockpit, the pilots and the visiting woman were riding down into the forest at last. The engine on the right was torn from its mount. As the airplane came to a stop, the right wing broke apart and fuel from its tank sprayed forward and ignited, causing an instant and intense fire just ahead of the final resting place and along the entire right side of the fuselage. Another fire, smaller, ignited at the root of the left wing, which remained attached.

For several seconds the cabin was silent. The emergency lighting had failed, and wires were shorting and sparking in the forward galley. For the moment everyone was alive. But fire penetrated from the infernos outside, and the cabin erupted with heavy black smoke and flames. On the left side the fire came through broken windows at rows 8 and 9. On the right side, where the whole world outside was burning, it came through a crack in the fuselage from rows 10 to 15. Because of superstition, there was no row 13, but this became the dividing line, a wall of flames and smoke that was impossible to see through, let alone cross and survive. People seated in rows 14 and behind rushed toward the rear, while people in rows 12 and forward moved in the opposite direction, toward the cockpit. Suddenly one woman's hair was on fire. Suddenly other people's clothes were burning.

The mood was something between urgency and panic. In the back the evacuation proceeded under the direction of a cool-headed flight attendant named Bruno Pichot, who several years earlier had helped to evacuate a burning Air France 747 that had crashed on takeoff in Mumbai, India. In the burning Airbus now, Pichot opened the aft door on the left side and deployed an inflatable slide into a tangle of broken trees outside. When the first passengers went out, their weight caused the slide to tear and deflate against sharp branches. Pichot dispatched a female flight attendant named Muriel Dager to help clear people out of the way at the bottom of the ruined slide, and he kept the evacuation going, using a combination of firmness and reassurance to overcome people's hesitation to jump. Some passengers were injured by their falls or by having other passengers fall on top of them. Inside the cabin, the fire was growing fast. After a man who appeared to be the last of the passengers made it out, Pichot escaped as well. But Muriel Dager was not convinced. She was thirty-one years old and extraordinarily brave. Despite her own injuries and what turned out to be deep emotional trauma, she climbed back into the burning fuselage to find anyone who might have been left behind. She intended to walk up the aisle checking row by row.

But by then the smoke and flames were so intense that she could do nothing but call into the inferno a few times, before jumping again to save her life.

All passengers in the back had escaped. But forward of the midcabin fire, toward the front of the airplane, the evacuation did not go so well. The smoke was thicker there, and obviously poisonous. People surged forward up the aisle, shoving ahead in a cluster. The senior flight attendant, a middle-aged man, was struggling with the forward door, which had opened partly but was blocked by branches outside. Complicating his task, the emergency slide had partially inflated into the cabin. The door to the cockpit was ajar, and the center console could be seen sparking. The copilot had been stunned by a blow to the head, and Asseline, who had also been injured, was trying to help him out of his seat. The visiting flight attendant emerged to join the struggle with the cabin door. When a passenger added his weight to the effort, the door suddenly opened, and both he and the senior flight attendant lost their balance and tumbled out. The slide then deployed on top of them. The visiting flight attendant stood at the doorway helping people onto the slide in rapid succession. None of them needed encouragement. But because the slide delivered them into a jumble of broken branches, they could not move away fast enough and began to pile into an impossible scrum. Observing the scene from the doorway, other passengers began jumping directly into the forest, but they, too, began to pile up. The senior flight attendant crawled out from under the slide and struggled to sort out the mess. Up in the doorway, the visiting flight attendant briefly paused the evacuation to allow the escape routes to clear. Overcome by smoke, she herself then went down the slide.

She was replaced by another flight attendant, who had been propelled forward by the crowd, but who managed to extricate herself in time to keep from getting ejected overboard. She stood in the doorway helping people to escape until no more appeared in the smoke. She shouted repeatedly for anyone left behind, and got no response. Asseline by now had seen to the copilot's evacuation, and he ordered the flight attendant to leave. She did. Asseline was certainly no coward. He felt his way back into the cockpit, hoping to put on a smoke mask and make a final inspection of the cabin. The fire, however, had grown so intense that he lacked the time and had no choice but to escape down the slide.

The episode should have ended there, with everyone having survived. Though some people had suffered broken bones, and a few had been severely burned, most had escaped with minor cuts and bruises. It actually was quite amazing. Of the 136 people aboard, only 34 required hospitalization, and 20, rather than waiting for rescuers, simply walked to the autoroute through the forest and hitchhiked home. Walking away was a cultural impulse. The French are less weak than they may seem, and they are also quite good at shrugging.

But for all the successful escapes in the end, it turned out that three passengers had died—each in the forward part of the airplane. One was a fourteen-year-old quadriplegic boy who had been paralyzed several years before in a car crash that had killed his mother. He was strapped into seat 4F, at a window on the right side. His father had offered him this ride as a special gift to allow him a normal life experience. At the time of the crash the father was waiting at the commercial airport for the flight's return from the tour of Mont Blanc. Seated beside the boy in the airplane was a friend of about the same age, who lacked the strength to lift him and got swept away in the evacuation. Amid the confusion, no one noticed the crippled child, who died alone in his seat. The second victim was a seven-year-old girl. She and her young brother

boarded the airplane late and were unable to sit together. Both
were given seats on the aisle, she in row 8, he a few rows behind.
After the crash she stayed in place, half-buried under a table tray
and unable to unlatch her seat belt. The strangers seated beside
her made good their own escapes and simply abandoned her to
fate. Her brother came down the aisle and struggled to free her,
but he was carried away by the crowd. Once outside the wreckage, he pleaded with others to return for her, but climbing against
the human flow would have been impossible, and by the time the
last person emerged—Asseline—there was no question of enduring the conditions inside. The third victim was a middle-aged
woman accompanied by her husband. They got to the forward
doorway, but as he escaped, she noticed the plight of the little girl
and plunged back into the cabin to save her. No one saw her do
this, but it is the only possible explanation. The woman and girl
died side by side.

•

No sooner had he walked away from the wreckage than Asseline
started blaming the airplane for what had gone wrong. It was a
position he maintained for years, even as he lost his pilot's license,
was convicted of manslaughter, spent time in prison, and moved
to Australia to resume his flying career for a local airline. The engines had failed to respond to the throttles, the computers had
assumed command, the Airbus had decided to land. The evidence
was all to the contrary and quite simple to understand. The airplane had performed exactly as it was supposed to. Moreover, it
had contributed significantly to the survival of the occupants by
refusing to stall even as Asseline flew it with the sidestick fully
back and sent it sailing into the trees. But Asseline would have
none of it now. When Ziegler spoke to him soon after the acci-

dent, he said, "Bernard, you know what a good pilot I am! It is impossible! I cannot have done this thing!"

Significantly, his copilot refused to comment—not then, or ever since. But the French union rallied to Asseline's side and drummed up support for his story among pilots across Europe and the United States. Claims were made that the official investigation was rigged, that the data contained in the airplane's "black boxes" had been doctored, and that the government was scapegoating Asseline to keep the Airbus company alive. The thinking was delusional to the extent it was believed. It did provoke one French 747 captain into becoming so obsessed with the case that he was committed to an asylum, and it continues, to a lesser degree, to inspire some conspiracists to agitate even today for justice. To such people the facts in this case have never much mattered. The accident presented an opportunity to fight back against Airbus and its fly-by-wire philosophies. As with every Airbus accident since, it was played out in public as a concern over safety, when it was really about ceding authority to machines, and the inexorable decline of a once-proud profession.

Part Three

SURVIVAL

EIGHT

THE DECISION

3:29 p.m.

Pilots don't get to make the purchase decisions. They fly the airplanes that their employers provide, and then they go home to their moonlighting jobs. That's the way it is. That's the way it will be unless the airlines, as long ago, are allowed to establish route monopolies. There is little chance of that happening, given the rise in ticket prices that would result. The trend is all in the opposite direction, toward the cheapest forms of air transportation. Opposition to Ziegler's control system philosophy has never been enough to overcome the savings offered by the overall designs. Far from it. The real choices are ultimately about money. By the time Sullenberger and Skiles took off from LaGuardia in early 2009, nearly five thousand fly-by-wire Airbuses had been delivered, and several thousand others were on order. Furthermore, after its acquisition by America West, US Airways had become the largest Airbus operator in the world, with more than two hundred fly-by-wire variants in its fleet, and it had stopped buying Boeings altogether. In a vicious twist during the long assault on its employees' salaries, the former management team had even demanded concessions at the pilots' expense in order, explicitly, to go shopping in Toulouse. When Sullenberger complained to Congress about being used as a cash machine, the unstated back-ground was that some of that cash had been spent to acquire airplanes that diminish the authority of pilots in flight. Was Sul-lenberger thinking about that? Certainly he knew the history. And certainly the union treated him as a godsend after his water land-ing. For some reason he has never commented on the Airbus de-sign. You can infer what you want from his silence, but you might easily be wrong. He is a careful man by nature. His opinion of the airplane remains unknown.

What is known, however, is that he and Skiles flew by wire during the glide. They had no choice. Like it or not, Ziegler reached out across the years and cradled them all the way to the water. His assistance may have been unnecessary, given the spe-cial qualities of these particular two, but there is no question that the practical effects were profound. At the moment of the bird strike, when the engines lost thrust, a conventional airplane would have tried immediately to nose down. It would have wanted to go into a sharp descent, and would have required whoever was flying to haul back on the controls with some strength and to retrim the airplane for a slower, more moderate glide, while disciplining the wings to stay level until the decision could be made to turn around. None of this is inherently difficult, but it imposes insidious de-mands on the crew in an emergency, when they are already busy with more important concerns. It is an accepted reality that the repetitive and menial jobs associated with baseline control subtly impinge on a pilot's capacities, and that during periods of truly high workloads, even simple thoughts are difficult to have.

Imagine trying to disarm a bomb while also having to deal with menial chores and talk on the phone at the same time. Sullenberger and Skiles disarmed a bomb on a three-minute fuse. They did it by concentrating on the two really important matters—how to get the engines started, and where to land. They could have done it in a Boeing, too. But it was helpful to their immediate cause that they were working with the product of Ziegler's mind, in which computers took care of the menial chores, then conjured up a magic carpet for them to fly. Even before the bird strike occurred, the ride was miraculous. Skiles was steering with his sidestick control, nudging it occasionally to make flight path adjustments, but otherwise leaving the airplane largely alone. The black box called the flight data recorder shows the functioning of the system. With the sidestick in its neutral straight-up position, the computers maintained the airplane in its last requested condition—in this case, with its wings completely level and its nose pitched 10 degrees high for the climb. With no pilot input required, the control surfaces out on the wings and tail were moving tirelessly to suppress any wobbles. When the birds then hit, and the engines lost thrust, the control surfaces had to work a little bit harder, but the nose stayed high in position and the wings remained perfectly level. Because the nose did not dip, the airplane decelerated strongly, but it stayed automatically in perfect trim, and it continued to climb, lofted by its mass to just over 3,000 feet. Had the pilots not intervened, it would have continued to decelerate straight ahead, retrimming itself until reaching the first of Ziegler's low-speed protections, at which point it would have gently lowered its nose and accepted a descent to maintain the margin above the stall. Given the airplane's weight, and with its flaps and slats retracted, this would have happened at about 170 miles per hour.

Sullenberger did not let the airplane slow to such a degree, because the wings at that speed would have plowed through the air and resulted in a sink rate that was unacceptably high. Twelve seconds after the bird strike, having switched on the engine ignition and started the auxiliary power unit in the tail, he assumed control from Skiles and nudged his sidestick forward to gain some speed, while simultaneously rolling into the left turn back toward LaGuardia. The climb crested as the nose eased down, and the speed began to increase above 213 miles per hour. Sullenberger's target was a green dot on his airspeed scale that indicated the best value to fly for maximum gliding distance. It was calculated by Ziegler's blessed system, and floated around 250 miles per hour. Sullenberger accelerated gently toward the speed, and stuck to it when he arrived, about halfway through the turn. It was a beautiful piece of flying. The control forces were light. Sullenberger was relying heavily on the automation, reversing course with a moderate 33-degree bank—the maximum angle at which the control system packages entire turns, automatically maintaining the bank and pitch angles, guaranteeing spiral stability, and, in this event, holding the best gliding speed with few corrections required. These specific flight characteristics are significantly different from those of conventional airplanes and are helpful to any pilot—even Sullenberger—during an emergency turn. The computers were extracting maximum performance from the wings and constantly adjusting the control surfaces to provide an extraordinarily stable platform to fly. As a result, though Sullenberger's workload was high, he was able to focus almost entirely on more important challenges than basic control.

They were still turning. Skiles continued down the dual-engine-failure checklist, one item at a time. He said, "ATC, notify. Squawk seventy-seven hundred." These were formalities, but he was trained not to skip steps.

Sullenberger said, "Yeah." His role was to be bottom-line. Speaking of the engine, he said, "The left one's coming back up a little bit."

Skiles said, "Distress message—transmit. We did."

They came out of the turn flying southwest on a heading of

230 degrees, and descending through 1,450 feet doing 240 miles per hour, slightly slower than the green-dot gliding speed. The nose was 5 degrees up. The wind at their altitude was blowing at about 20 miles per hour from the right and slightly from behind. The Hudson River stretched ahead, beckoning from the far side of the George Washington Bridge. The surface looked smooth. LaGuardia Airport was now visible at a sharp angle to the left. The threshold to Runway 13 lay six miles away. This was the closest runway to the airplane—the one that the controller Patrick Harten had just offered and that Sullenberger had rejected during the turn. Of course it was possible to go for it anyway. As an experienced pilot, Sullenberger was skilled at visually projecting descent paths, even around corners. Now that the airport was in sight, it was not obvious that if they headed directly to the runway they would fail to reach it. But at some point they would have to extend the landing gear, slats, and flaps, and this would increase the descent angle to an extent not visible. The airplane did not seem to have the margins of altitude that would allow Sullenberger to adjust the flight path to navigate a surefire approach to a safe landing. He decided again not to try for the runway.

Months later, in the more leisurely environment of an Airbus simulation center in Toulouse, the NTSB ran a study of the choices available. A simulator was programmed to duplicate the circumstances of Sullenberger's bird strike—weight, temperature, speed, altitude, winds, location, some slight thrust from the left engine—and four pilots were enlisted to fly a series of attempts on LaGuardia. In the setup there were two important differences from the actual flight. First, the starting point was the location of the bird strike itself, not the location where Sullenberger came out of his turn. Second, the pilots knew the game in advance. Being aware turned out to be crucial. In every case where the pilots were allowed to respond immediately to the loss of thrust by making a quick turn back to the airport, every one of them was able to land safely. The quick turns were flown at steep bank angles—presumably up to the maximum of 67 degrees allowed by the system—while relying on fly-by-wire protections against stalls. Since all of these pilots succeeded, it is likely that Sullenberger would have succeeded as well. But how realistic is it to expect any pilot to have responded immediately to the surprise over the Bronx and to have hooked a high-G turn with a planeload of average passengers in the back? Sullenberger required eighteen seconds from the bird strike to the start of his more moderate turn, and it is hard to imagine anyone performing better. In recognition, the NTSB then imposed a thirty-second delay before allowing the simulator pilots to fly their returns—and every one of them crashed.

Sullenberger made the right decision. No matter what. Even if every simulator run had later succeeded. Even if people had died because of the landing in the Hudson. Going for Runway 13 would have been a crapshoot in an environment of obstacle-strewn waters, where missing the runway by fifty feet is like missing it by a mile. Who knows what Piché might have tried? But Sullenberger had never been that kind of gambler.

When Sullenberger rejected the offer of Runway 13, Harten correctly made no assumptions about why. For all he knew, the airplane had been damaged to the extent of being difficult to maneuver sharply, and in conjunction with that was perhaps even too high for the nearest threshold. Now, as the flight came out of its turn, he offered up the same runway in the opposite direction. Harten radioed, "All right, Cactus 1549, it's going to be left traffic to Runway 31." If needed, Sullenberger could overfly the airport, then turn around and come back in.

This was out of the question. Sullenberger answered with a terse "Unable."

In the cockpit, a conflict alert sounded: *Traffic. Traffic.* It was junk in the context. Neither Sullenberger nor Skiles made any comment.

Harten radioed, "Okay, what do you need to land?"

Total engine failure imposes surprise destinations. If not Charlotte, then some other airport. If not an airport, then an unobstructed highway, or, in descending order, a large flat field, an especially large golf course, a forest, or in the extreme, a lake or river for a ditching close to shore. At some point, as you climb down from the most desirable destinations, you stop thinking about where you might spend the night, stop thinking much even about the airplane, and shift your focus to survival. At that point life becomes very simple. The first rule is to avoid losing control. The second rule is to avoid hitting brick walls. The third and final rule is to keep flying the airplane even as it is sliding and disintegrating around you in the water or on the ground. You fly it until it stops, and then you evacuate.

Sullenberger did not answer Harten's question. He was still looking for better solutions, but beginning to set up for the Hudson. Talking on the radio is low on the list of chores. You fly the airplane first, you navigate second, you talk on the radio after that. Sullenberger was clear about the priorities. His silences were brilliant.

Skiles was perhaps unsure about them. He hardly knew Sullenberger. He had gained a good impression of him in ordinary flying, but he did not know how he would react in a crisis. Skiles mentioned Harten's offer. He said, "He wants us to come in and land on [Runway] 13 . . . For whatever . . ."

The cockpit sounded with another junk alert: *Go around. Wind-shear ahead.*

Skiles got back to the checklist. "FAC One—off, then on."

Harten kept offering choices. "Cactus 1549, Runway 4 is available. If you want to make left traffic to Runway 4."

This time Sullenberger answered. "I'm not sure we can make any runway. Uh, what's over to our right? Anything in New Jersey? Maybe Teterboro?"

He was speaking faster than he had before. The airplane had just overflown the Manhattan side of the George Washington Bridge and was gliding through 1,200 feet. From the glass-walled cab of the LaGuardia Tower, controllers were following it with binoculars. The pilots of two helicopters flying Hudson River tours spotted it as well.

Harten was the observer with the most important view. He still had clear communications and good radar returns. He radioed, "Okay, yeah, off to your right side is Teterboro Airport."

Sullenberger was silent. The conflict alert announced: *Monitor vertical speed*. Junk. Sullenberger raised the nose slightly, and the airplane responded by coasting nearly level at 1,100 feet, as the speed bled off to 220 miles per hour—30 miles per hour slower than the green-dot best gliding speed. During his interviews with investigators after the accident, Sullenberger did not seem to remember that he had slowed in this manner, but he had, and it was an excellent move. The green-dot speed delivers the longest glide as measured in distance, but not as measured in time. The latter is delivered by a slower speed known as minimum sink. In the A320 it is close to the speed that Sullenberger was now flying. With the bridge passing beneath them and the river stretching from directly below to several miles ahead, distance no longer mattered, but any additional time that could be gained would increase the chance of restarting at least one of the engines.

Skiles was just now coming to that part of the checklist. He said, "[If] no relight after thirty seconds, engine master [switches] one and two confirm . . ."

Harten radioed, "You wanna try to go to Teterboro?"

Sullenberger answered, "Yes," but only because he was considering possibilities. Teterboro is an airport surrounded by city. It was no closer than LaGuardia. They were too low and slow to get there, and he indicated no move in that direction. They remained over the Hudson, descending.

Skiles continued: ". . . off."

The conflict alert announced: *Clear of conflict.*

Referring to the engine master switches, Sullenberger repeated, "Off."

The restart procedure called for both engines to be switched off simultaneously, and, after a delay of thirty seconds, restarted one by one. Skiles was working the switches. He knew that the left engine was still producing electrical power, and perhaps some small amount of thrust, and he balked at shutting it down until first trying the right engine, which was contributing nothing at all. Sullenberger implicitly agreed. If it ain't broke, don't fix it; if it is broke, you've got nothing to lose from trying. Skiles switched off the right engine only. Reading from the checklist, he said, "Wait thirty seconds."

Sullenberger made his announcement to the cabin. He said, "This is the captain. Brace for impact."

The flight attendants started shouting, "Brace! Brace! Heads down! Stay down!" Because they could not see outside, they did not know that the airplane was over the Hudson and headed for the water. Their ignorance did not matter: an impact was imminent, and they would hardly have taken this moment to inform the passengers of the life vests under their seats and to order the life vests put on.

Two minutes had passed since the encounter with the geese. In the cockpit a ground-proximity warning announced the altitude in feet: *One thousand.*

Skiles wanted to jump ahead of the thirty-second delay in attempting to restart the right engine. He suggested, "Engine master two, back on . . ."

Sullenberger agreed. "Back on."

Skiles flipped the switch. He said, "On."

Harten had spent the previous few seconds on the phone with Teterboro Tower, and with the utmost brevity had arranged for the emergency arrival. Now he radioed, "Cactus 1529, turn right 280 [degrees], you can land Runway 1 at Teterboro."

Rather than springing back to life, the right engine responded to the restart procedure by grinding nearly to a halt. Looking at the ultra-low indications in the cockpit, and apparently addressing the engine itself, Skiles said, "Is that all the power you got?" He was not imploring the engine as Asseline's copilot had. He was probably a little disgusted.

It was time to turn his attention to the left engine, known to pilots as "Number One." Skiles hesitated. He said, "Wanna . . . Number One? Or we got power on Number One." He wanted confirmation from Sullenberger.

Before answering Skiles, Sullenberger got back to Harten on the Teterboro offer. He radioed, "We can't do it." Then he said, to Skiles, "Go ahead, try Number One." However helpful it had been during the glide, that engine was not producing nearly enough thrust for level flight. With the right engine now conclusively dead, the near-certainty of a water landing ahead, and no need to stretch the glide, logic required that restarting the left engine be tried.

Skiles flipped the master switch to "off" as required, and the left engine immediately spooled down, with its propulsion fan slowing to a miserable 15 percent.

Meanwhile, in his radar room at New York Approach, Harten wasn't going to give up on these guys. He believed that they and their passengers were about to die. He radioed, "Okay, which runway would you like at Teterboro?"

Sullenberger answered, "We're gonna be in the Hudson."

"I'm sorry, say again, Cactus?"

Cactus did not answer. Harten kept after his other duties. He radioed, "Jetlink 2760, contact New York Center, [frequency] 126.8."

"Twenty-six eight, Jetlink 2760." It is almost certain that those pilots kept a radio tuned to the unfolding drama.

In their cockpit, Sullenberger and Skiles were down to 650 feet, and doing 218 miles per hour. This was not the moment for the thirty-second delay. Nine seconds after switching off the left engine Skiles said, "I put it back on."

Sullenberger welcomed the move. "Okay, put it back on, put it back on." He was focused on the water ahead. The fly-by-wire system was continuing to work closely with him, making constant adjustments to the control surfaces to keep the airplane absolutely steady and responding smoothly to the sidestick movements. The Airbus glided down through 400 feet. The river ahead was wide. It was clear of boat traffic. Sullenberger eased slightly from the New York to the New Jersey side of the center stream. In the cockpit there was little sensation of speed. A ground warning began to repeat: *Too low. Terrain. Too low. Terrain.*

Sullenberger and Skiles were necessarily concentrated, but not unnaturally cool. Sullenberger was doing the basic work. Having switched on the left engine again, Skiles was watching the indications. The engine returned to life, but barely, not even to the extent of before. After about eight seconds Skiles concluded the worst. He said, "No relight." It was a simple statement of fact. Two words. The end had come. They could not wish away the water.

Sullenberger said, "Okay, let's go. Put the flaps out, put the flaps out."

The box announced, *Caution. Terrain. Caution. Terrain.*

Skiles said, "Flaps out?"

Yes.

Terrain. Terrain. Pull up. Pull up.

They could have followed the advice. Sullenberger could have snatched the stick as Ziegler envisioned, and relied on the fly-by-wire protections to zoom the airplane up with no fear of exceeding the structural limits. He could have spent all of the remaining airspeed on doing that climb—paid every last dime of it out. But then, with no thrust to apply, the airplane would have hung high on the edge of a stall and, on the far side of the climb's crest, started down, nose up, wings plowing, with no speed left to spend to keep from hitting hard. Water hit hard is as hard as concrete. So, no, he was not confronted with a mountain in Cali. He was not that kind of pilot. Through no fault of his own he was committed to a landing in the Hudson River, and he needed to make it smooth.

Machines can be so dumb. The warning system continued to nag. It kept insisting, *Pull up, pull up*, over and over again. The pilots could have muted the alarm—somewhere the checklist called for this—but they were busy with the job at hand. Skiles slid the flap lever to position two, which extended the wings' leading-edge slats to the nearly full position and dropped the trailing-edge flaps to half. With the curvature provided by these devices there was some additional drag, but the wings had been reshaped to bend the air better and continue to lift the airplane even as it slowed.

Harten lost sight of the airplane on his radar screen as it dropped below the skyline. He did not know if the pilots could hear him anymore, but he offered them choices in case they could still get an engine going. "Cactus 1549, radar contact is lost. You . . . also got Newark Airport off your two o'clock, in about seven miles."

The transmission came clearly over the Airbus radio. Sullenberger ignored it. Skiles said, "Got flaps out." Then he said, "Two hundred fifty feet in the air."

Too low. Terrain. Too low. Gear.

Harten handled another flight: "Eagle Flight 4718, turn left, heading two-one-zero."

The pilot answered, "Two-one-zero, ah, 4718. I dunno, I think he said he's goin' in the Hudson."

In Flight 1549 the flight attendants had switched to "Grab ankles! Heads down! Stay down!" The airplane was slowing below 200 miles per hour. Skiles said, "Hundred and seventy knots." With nothing to lose by trying, Sullenberger advanced the throttle for left engine. Skiles said, "Got no power on either one? Try the other one."

Sullenberger said, "Try the other one." He did. There was nothing.

Harten radioed, "Cactus 1549, uh, you still on?"

Caution. Terrain.

Skiles said, "Hundred and fifty knots . . . Got flaps two, you want more?"

"No, let's stay at two."

Caution. Terrain.

Sullenberger said, "Got any ideas?"

Skiles said, "Actually not."

NINE

THE FLARE

3:30 p.m.

In the annals of jet transport intentional water landings—ditchings—are extremely rare. Airliners do go into the water with some frequency, but out of control and without planning, usually staggering or slipping off the far end of a runway after a failed takeoff or a landing gone wrong. US Airways has done it twice at LaGuardia alone. Though the impact forces may break the airplanes apart, these accidents have turned out to be survivable for most of the people aboard. Beyond that, because of the chaos that results from loss of control, few generalizations can be made. Post-crash fires are rare. Impact forces may kill some people. Drowning may kill others. The pilots become helpless occupants, usually responsible for the disaster, but finally just along for the ride.

Not so for Sullenberger now. However short the notice had been—about three minutes and counting—he would hit the water in an airplane under control, in a crash that could be planned. The planning had started years before, at the time of the airplane's design, in order to meet European and American certification standards, which require manufacturers to demonstrate that their airplanes can be ditched. These are paper proofs, not physical demonstrations. They are founded on a body of knowledge gained during World War II, when thousands of crews ditched stricken airplanes at sea. The experience was supplemented by systematic research in the 1950s, using scale models in water tanks. About the dynamics of splashdowns, therefore, some basic truths are known. Speed is bad. Size is good. A low-set wing is better than a high-set wing, but the positioning should allow some belly below. The rear of the fuselage always hits first. A flat belly there helps the airplane to skim along the surface and maintain a decent angle as the airplane sinks in. If the landing gear is down, it may cause the nose to plunge violently beneath the water. This is known as submarining. The effect is like hitting a wall. For ditchings, therefore, the gear remains retracted. The belly bears the brunt of the contact. The skin may bend, but must not break. If it breaks, water jets in with terrible force and eats out the airplane's insides. This is not ideal for survival. Ideally the flaps should be lowered before touchdown, but only if they can be counted on to give upward on impact. Low-slung jet engines add complications. Ideally they will shear off rather than destroy the wings and rupture the tanks.

Then the airplane comes to a rest. It lies low in the water. For argument's sake during the certification process, its fuselage is believed to have remained whole, and perhaps without engines, but with the wings still attached. If it has ditched into an ocean, the standards allow waves to have threatened the airplane during the last seconds of flight—and expect the pilots to have come in parallel to those waves, banking left and right to match the watery slopes, to keep the wingtips from dragging. Now that the maneuver has succeeded, however, the waves are regulated away; for argument's sake, suddenly the water is flat and the wind is calm. An airplane in those conditions will float nose high on an even keel, settle in a predictable manner, and disappear tail first below the surface. The timing can be known based on weight, buoyancy, and leakiness. Certification demands that the angle of the aisle not be too steep for passengers to climb, and that the lowest exit sills remain above water long enough for all the occupants to escape, even if half the exits are not available. Airbus engineers equipped the A320 with a special ditching switch to close certain inlets below the theoretical waterline. They then showed that the lowest (aft) doorsills would remain above water for an ample seven minutes and fifteen seconds. This required that conditions would cooperate and that the airplane would be well flown through the landing. The accepted scenario allowed the theoretical pilots to come in with theoretical power, on the assumption that the theoretical ditching was not the result of a dual-engine failure, but rather was a precautionary water landing for reasons not specified. The A320 would touch with the gear retracted, using full flaps, with the nose pitched 11 degrees up, on a nearly flat approach to the water, feeling for the surface gingerly by 3.5 feet per second, at 136 miles per hour. Airbus proved mathematically that the belly had the strength to handle the hit. And it was obvious the maneuver would be possible to fly, because it is almost the same as that of a normal landing on land. In theory, the profile could even be flown after a dual-engine failure, but the subject was hardly addressed. Airbus was answering to certification standards, not looking for problems in the abstract.

The intentional ditching of a jet airliner? There is a shadowy story from the Soviet Union about a couple of Aeroflot pilots opting for the river in Leningrad in 1962 after engine failures in a new Tupolev design, and making a landing that everyone survived. The story was suppressed because of the government's concern for the reputation of the airplane type, and it emerged only decades later, after the fall of the USSR. Meanwhile, in 1970, an American crew flying for an obscure airline ditched a DC-9 into the Caribbean. They had flown from New York to Saint Maarten, had burned through their fuel reserves by botching multiple approaches in marginal weather, and were trying to divert to Saint Croix when they realized that they were not going to make it. Though there was ample time to prepare the cabin for the ditching, and the engines kept running through a systematic descent, the passengers were left largely on their own, and communication between the cockpit and cabin was so poor that one flight attendant was still standing in the aisle when the engines flamed out and the pilot flared down into the waves. There was some success. The airplane held together, and the fuselage, though torn, floated for seven minutes, allowing for the evacuation of those occupants still capable of movement. But of the sixty-three people aboard, twenty-three people died, most from blunt trauma. The dead included the flight attendant in the aisle, passengers whose seat belts were unfastened, passengers whose seat belts were fastened but failed, passengers whose entire seats were torn from the floor, two unsecured babies who never had a chance, and several people who escaped from the cabin but drowned. At least the water was warm. The airplane sank through 5,000 feet of it, with one of its life rafts half-inflated inside the galley and the others neatly stowed in their bins. On the surface, people clung to flotsam and an inflated slide to survive. The first were rescued by helicopters after drifting for an hour and a half. The ditching was a fiasco start to stop. It is hard to see what advantages the "planning" had brought. The passengers might have fared better had the pilots simply slid into the water off the far end of the Saint Maarten runway.

Afterward, no jet airliner was intentionally landed in the water for a very long while. An Eastern Airlines L-1011 came close to it above the Atlantic off Florida in 1983, after losing thrust from all three engines. During the anxious glide toward the coast, the flight attendants prepared the passengers for a ditching—but then the pilots got an engine started and proceeded to a safe landing in Miami. Pertaining to customer relations, some small lessons were learned. One is that nonswimmers especially may panic at the sight of life vests. Another is that once certain people start to scream they cannot be stopped. The last is that passengers should not be made to brace indefinitely while waiting for an impact to come.

But that was just a practice run. For thirteen years the scene was quiet. Then, in 1996, an Ethiopian Airlines Boeing 767 went into the water off a beach resort in the Comoros islands after being hijacked and running out of fuel. It hit at high speed with its left wing down, and came apart spectacularly as it spun around, leading to the death of 125 of the 175 people aboard. The accident was an anomaly, an intentional water landing, yes, but largely out of control because the pilots were fighting with the hijackers at the time. Some passengers drowned because they had inflated their life vests prematurely and were pinned against the ceiling as the cabin sank. Beyond that, as for ditchings, there was nothing to be learned. Five years later, in 2001, Robert Piché ran an extended exercise during his glide to the Azores—but of course, he ended up landing on a runway. The real thing didn't happen again until January 16, 2002, thirty-one years after the genuine affair in the Caribbean, when two Indonesian pilots flying a Boeing 737 with sixty people aboard lost thrust from both engines during a descent over Java in heavy rain and hail, and ditched into the shallows of the muddy Solo River. The airplane came to a stop in three feet of water, resting on the river bottom with the fuselage intact. One flight attendant died by drowning, but everyone else waded safely ashore. Locally it was quite an event. By the second day, villagers were selling access to crowds who arrived to admire the wreckage.

•

Seven years later, Sullenberger was next in line. Like other airlines, US Airways practiced its crews on the use of life vests and rafts, but offered almost no training on how to fly a water landing. The lack of specific training makes institutional sense, given the rarity of the event, the value of training time in other areas, the difficulty of building a simulator that can take waves and other variables into account, and the fact that the profile to be flown closely matches that of normal landing. Or officially it does. This goes back to the certification requirements. From the government regulators, to the Airbus factory, to US Airways, to the approved ditching checklist provided to pilots in the cockpit, the assumption is made that at least one engine is running. With that in mind there is plenty of time to tidy up the cabin, get the passengers into their life vests, push the special ditching button, and use power to ease the final descent into the perfect groove of a gear-up, full-flap, 11-degree-nose-high, maximum 3.5-feet-persecond, 136-mile-per-hour delight. If the groove isn't right on the first try, hell, you can probably even go around and try again. Never mind that no jet airline pilot has ever ditched an airplane as a precautionary measure at a time when the engines were still running. Not the Russian in Leningrad, or the American in the Caribbean, or the Ethiopian in the Comoros, or the Indonesian in Java—and certainly not Sullenberger now.

There was no time for the ditching checklist anyway. Even the dual-engine-failure checklist was an unwieldy tool. It assumed that the engines had failed at high altitude, and took a pedantic approach to solutions, starting with the advice "LAND

ASAP," followed soon afterward with "LANDING STRATEGY—
DETERMINE." In small print, it then whispered, "Determine
whether a runway can be reached, or the most appropriate place
for a forced landing/ditching." Oh, really? The checklist was three
pages long, and had to be taken in sequence. In the three and a
half minutes available, Skiles barely finished with page one, and
only then by skipping the standard delay for switching the engines
from off back to on. Had he been able to continue, on the third
page he would eventually have come to the section offering guidance on water landings. But by now they had descended through
400 feet, and there was nothing to do but fly.

Sullenberger was brilliant at it, as was the automation he com-
manded. Their roles were linked but distinct. His was to make the
decisions that mattered. The automation's was to execute them
well. Together now they would have only one chance to set the airplane down. Sullenberger had been thinking about it throughout
the glide. Ahead lay the problem of energy management, as he later
described it—the trade-off between speed and altitude. He had selected slats and half-flaps to allow for the slowest possible landing,
but had decided against the standard full-flap setting, because of
the associated drag. Without thrust he was approaching the surface
at a descent rate that already was abnormally high. For that reason
he was carrying some extra airspeed to expend on tapering the descent by raising the nose and flaring. He was familiar with the process from every landing he had ever done—but this time his depth
perception would be limited by the featurelessness of the forward
view. The airplane would be in an unfamiliar configuration. And he
would be managing an energy trade with values that were extreme.
To his advantage the river was long, and he could touch down any-
where on it. But the landing gear was up, the stakes were high, and
there would be no chance for a second try.

He started raising the nose at about 200 feet—around when Skiles confirmed that the flaps were set—and apparently he over-rotated very slightly, because the airplane ballooned into a shallow climb, squandering some of the energy that he had guarded. This is when Skiles read the speed, "One hundred fifty knots." Sullen-berger was completely in control. He eased the stick forward even as the climb crested and, mushing nose high, the airplane started down. Speed was suddenly in short supply. This is when Sullen-berger had the presence of mind to ask Skiles if he had ideas, and Skiles had the cool to say, "Actually not." The fluency they exhib-ited at such a critical moment, in continuing to discuss matters calmly, helps to explain why their passengers survived.

In the radar room at New York Approach, Patrick Harten be-lieved they had already died. Transmitting in the blind for one last time, he radioed, "Cactus 1529, if you can, uh . . . You got, uh, Runway 29 available at Newark. It'll be two o'clock and seven miles."

On the far side of its little climb, as the airplane sank toward the surface again, Sullenberger resisted with a delicate touch, bringing the sidestick back and parsing out the reserves of air-speed he still had left to spend. *Terrain, terrain. Pull up, pull up, pull up.* The warning repeated from now until the end. For fif-teen seconds the pilots flew in silent concentration as Sullen-berger finessed the flare. They were coming in at a steeper angle and higher descent rate than test pilots would later be able to achieve during forewarned risk-free simulations of the maneuver. In the actual Airbus, Ziegler's computers responded as instructed, inching the nose higher as the airplane descended and, with in-creasing active control surface movements, keeping the wings completely level according to Sullenberger's eyes. His eyes were good. The surface was close. The sidestick was largely back. Sullenberger said, "We're gonna brace." But they had to keep flying. They had five seconds left, with no guarantee of success.

•

Let others speculate about what could have been. What matters is what happened. Back at the moment of the engine failures, even before assuming control, Sullenberger had started the auxiliary power unit—the APU, in the tail, which provides triple-redundant electrical power by driving the airplane's third generator. There was no harm in starting the APU, but it was not a book procedure. Investigators later wondered why Sullenberger did it. In his interviews with the NTSB, he was not entirely clear: it had seemed like a good idea at the time. He could have said more. Across a lifetime of flying, Sullenberger had developed an intimacy with these machines that is difficult to convey. He did not sit in airplanes so much as put them on. He flew them in a profoundly integrated way, as an expression of himself. He lived through them. He knew their souls. And this is not speculation. It is a reality for thousands of working pilots who feel deeply at home in the sky. Call his act an inspiration, or find technical reasons if you must, but Sullenberger just knew that he needed to get the APU running. It's functioning was irrelevant at first. Though the generator driven by the right engine had dropped off-line from the start, the generator driven by the still-productive left engine core continued to provide plenty of charge all on its own. But two minutes into the descent, when Skiles tried to restart the left engine, the core speed plummeted, and never came back to more than 50 percent. The minimum speed at which the engine core drives the electrical generator is 54 percent. The effect of the restart attempt, therefore, was to drop the left generator off-line.

Had the APU not been running, the airplane would have reacted as the A330 did for Robert Piché during his glide to the Azores, by extending a ram air turbine windmill into the slipstream, to drive a small emergency generator. That generator produces only a fraction of the airplane's electrical needs. To preserve the batteries from exhaustion, the computers would therefore have staged a planned withdrawal, blanking out unessential instrumentation in the cockpit and reducing the operation of the fly-by-wire controls by one notch, from full-up "normal law" to what is known as "alternate law." By natural-law standards, normal law is normal only as Ziegler wished normal to be: it is the uncompromised fly-by-wire package, the magic carpet ride. Alternate law is an intermediate condition, halfway to a third stage, called direct law, where for lack of computational power, the Airbus behaves like a conventional airplane. The subject is intricate, and too tedious to delve into here. Suffice it to say that had the airplane dropped into alternate law, its flight envelope protections would have disappeared. This never happened, but only because Sullenberger had started the APU. Ziegler answered with normal law all the way to the river.

Deep into the flare, Sullenberger pulled into the zone of low-speed protections. This is to his credit, and a measure of the extraordinary last stage of the flight, with no thrust to cushion the descent, with limited depth perception over a river, and with an imminent inevitable need to belly the airplane in. Sullenberger intended to touch down with the nose 10 degrees up, and he had to get there at just the right moment, pulling up to the angle neither early nor late, in order to spend whatever last airspeed still remained to taper the airplane's closing rate on the river. He was not thinking about the certification standard (with engines running) of a maximum 3.5-feet-per-second sink rate at water contact, but he knew he could not afford to drop in hard. The flight data recorder bore witness to his work. Ten seconds before touch-down, the airplane was suddenly sinking much too fast, at a rate six times higher than the belly could endure. He lowered the rate by raising the nose. Five seconds later, as he said, "We're gonna brace," he had split the difference in half, but by pulling early to 10 degrees of nose-up pitch.

It was a difficult balance. Skiles said something about a switch. Sullenberger said, "Yes." He had no choice but to keep raising the nose.

A ground warning announced thirty feet.

The airplane streaked into view of the shore-based security cameras upriver from the USS *Intrepid* museum. At that moment Sullenberger ran out of options. The nose was 11 degrees up, and the sink was still a bit too strong, but he had brought the sidestick fully back and had no speed left to spend. Energy management. He had come to Ziegler's limit: alpha max. Given the circum-stances, his timing was astonishing and almost perfect. Years in the future, when he looks back, he may regret having chosen an airline career, but he will have proof in the data from this flare that he was a pilot at the peak of human performance. The fly-by-wire system took it from there. It was a very brief affair. Because of the airplane's inertia, it was probably unnecessary. But for the last few seconds of the glide, with Sullenberger's stick fully back, the computers intervened and gently lowered the nose to keep the wings flying.

One of Ziegler's test pilots made a point to me over a slow replay of the effect. Asseline and Sullenberger could not be more different as men, and their accidents represent opposite extremes of the piloting spectrum, but there is striking similarity in the footage of the final moments of their accidents, with both air-planes hitting their tails first and crashing almost gently, with their

wings lifting at alpha max and the fly-by-wire system still working
to maintain control and avoid stalling. Asseline had the advantage
of settling through breakaway branches, with his landing gear extended and his engines producing thrust. Sullenberger had the
disadvantage of settling harder, gear up, and coming down onto
concrete. When the tail touched the water, the nose was pointed
9.5 degrees up and the airplane was flying at 144 miles per hour.
It was sinking 13 feet per second, more than three times faster
than the belly was built to withstand. As a result, the belly ripped
open at the point of contact, aft, and a destructive plume of water
penetrated the fuselage, tearing up structure and rupturing the
cabin's rear pressure bulkhead, or wall, through which the river
began to flood. Simultaneously the lower luggage area was crushed
upward, and a jagged beam penetrated the cabin floor perilously
close to the jump seat where the veteran flight attendant Doreen
Welsh was stationed. With the tail firmly scooping the river, the
wings came down, the left engine tore away, and the airplane slid
forward, skimming the surface in a veil of spray before skewing
left, tipping right, and coming to a stop. Sullenberger and Skiles
looked at each other in surprise. They agreed that the ride had
been less rough than they had expected. In other words, they were
alive. But it was not over yet. In the cabin sat three flight attendants and a hundred and fifty passengers. Outside lay the Hudson
in January. Sullenberger called for an immediate evacuation.

TEN

THE ESCAPE

3:31 p.m.

The cabin had twelve first-class and 138 economy-class seats, arranged in twenty-six rows. It had four main exit doors, two in the front and two in the rear. The doors had inflatable slides that could be detached and used as rafts. At rows 10 and 11, there were also four emergency over-wing exits, two on each side. Opening them automatically inflated one small slide along the fuselage, behind each wing. The wing slides were not detachable, and were not meant to be rafts. After the crash, this was the geography of survival.

As usual, some passengers were anxious from the start. One of them was an experienced flier, but also a new mother with a nine-month-old child at home. She sat at a window seat in row 20, six rows from the back, and at liftoff began counting the seconds, because she had heard that the first two minutes of a flight are the most dangerous. This is indeed true if you want to measure risk second by second, because the initial climb is over so quickly that it constitutes at most 1 percent of a typical flight, but accounts for 8 percent of fatal accidents in commercial jets worldwide. If you include the takeoff roll in your worries, you can bump the cumulative fractions respectively to 2 and 20 percent. That leaves 98 percent of the flight time ahead, as well as the segments in which 80 percent of fatal accidents occur—especially during approach and landing.

But passengers do what they can to endure the misery of airline travel, and while others in the cabin went catatonic, this woman got her one-second intervals almost exactly right. Who needs flight data recorders when there are anxious mothers aboard? One thousand one . . . two . . . three. She nearly made it to two minutes, but on the count of ninety the airplane hit the birds. She later told investigators that she heard a boom "similar to something being sucked into a vacuum cleaner." Like geese. She was a good observer. She did not see the brown flash of birds going into the engines, as others did, because she was sitting behind the wing. But she did see the fireballs of the compressor stalls and realized that the right engine was badly damaged. She felt the airplane wiggle in surprise. She heard a collective gasp from the passengers. She did not know that both engines were out. Quite reasonably, she assumed that the airplane would be heading back to LaGuardia. Entering the water did not cross her mind.

Sitting in the row in front of her was another anxious mother—the woman with the baby boy on her lap. Before the departure, she had been protesting the separation from her husband and daughter, who were seated three rows back and on the far side, but a flight attendant—probably Doreen Welsh—had settled her down. The woman had a middle seat, with her baby. The man on her left was sleeping. The man on her right coo-chi-cooed the baby, and informed her that he was the father of five. That seemed to make her more comfortable. He was a tall, athletic man, exuding self-confidence and calm. When the birds hit, he said, "Oops, that's an engine problem," and assured her that they would be okay.

The 293-pound woman was sitting three rows back, in 22D. The father and daughter were across the aisle and behind her, in 23A and B. On the right side, in 23E, was a young man in glasses, who would end up stripping to his underwear to swim. In 24A was a man who would end up hitting his head on his computer in the seat pocket in front of him, but without suffering bruises. Directly behind him, in 25A, was a California businessman who six months later was the only passenger to testify at the NTSB hearing in Washington.

The above is just a sampling. The crash investigators later interviewed nearly everyone aboard and summarized their experiences for the record. Altogether forty-nine people sat in rows 19 and higher, including the lap baby and Welsh in her jump seat farther aft. As a general rule such positions are the safest in a crash, because they are the farthest away from the initial point of impact, which is usually the cockpit. Pilots, it is said, are always the first to arrive at an accident, and they are more likely than others to pay with their lives. But the rule is reversed during controlled water landings, when the rear of the fuselage hits first and hardest, and airplanes, after coming to rest, sink tail low as they drown. This certainly seemed to be the case here, where proximity to the tail correlated approximately to the intensity of the experience, and Doreen Welsh, seated farthest back, had the hardest time of all.

Before touchdown, however, the experience was roughly the same. After the bird strike the ride remained airline-style smooth. People heard the hit, felt the deceleration, smelled the smoke, and realized that the airplane was turning, but they did not know that both engines were out—and in many cases would not have believed that an airplane could fly so well without power. A Spanish-speaking woman who was sitting with her six-year-old daughter in the front row of the economy section gave the only discordant description of that stage. She said that she had felt that something

was wrong even before the departure, that she had waited for the
flight attendants' safety demonstration but did not remember
hearing it, that after the bird strike the airplane turned quickly,
and that passengers were screaming, "Pray! Pray! Pray!" This appears to be a Latin version of events. Though a few people did
quietly pray, a larger number reached for their mobile phones. A
woman got through to her husband. The connection broke. He
called her back. Another woman sent a text to her husband, reading "My flight is crashing." Later she explained: "I didn't want him
to have three or four hours of this ambiguity—was I on that flight,
was I not? I wanted him to be able to get to the business of telling
our children and moving on." If this seems a bit rushed, there
wasn't much chance for contemplation. What almost all the passengers later agreed on is that perceptions of time did not slow
but accelerated, and that a stunned silence prevailed in the cabin
during the glide. Some people held their neighbors' hands. Some
murmured reassuring words. But it grew so quiet in the cabin
that as far back as the economy section, muffled cockpit warnings
could be heard. *Terrain, terrain, pull up, pull up.*

Two minutes after the engines quit, and one and a half minutes before the tail hit the water, Sullenberger ordered the cabin
to brace for impact. The order came as a surprise to nearly
everyone. One man said out loud, "What does that mean?" Soon
enough he figured it out. A woman sitting at the window seat 18A
reacted by closing the shade. The man sitting beside her asked
her to raise it again so that he could see outside. Apparently he
was polite, and she complied. The most astute passengers had
known for a while that they were descending over the Hudson, and
would not be returning to LaGuardia, but some had held out hope
that they were headed for Newark instead. Now they knew that

the airplane was going to crash into the river. The flight attendants did not know it, because in a lapse of aircraft design, they had no eye-level windows while seated in their positions, and were expected to rely on instructions from the cockpit at a time when, obviously, the pilots would be busy simply doing the flying. They therefore reacted purely by rote, chanting, "Brace! Brace! Heads down! Stay down!" with no idea of how high they were, where they were, or what was going on.

When instructed to brace, the passengers assumed all sorts of positions, from full head-down ankle grabs, to cross-arm bends against the seatbacks in front of them, to variations that allowed them to see outside, to nothing at all because they were frozen in panic or resigned to the idea that to crash in an airplane is necessarily to die. This last idea is widespread in the public because of media coverage that concentrates on the worst cases—TWA 800, ValuJet 592, United 93—but the NTSB found in a study of the most lethal 50 percent of U.S. airline accidents from 1983 through 2000 that 86 percent of the occupants survived. With the active engagement of Ziegler and his Airbus engineers, Sullenberger and Skiles were about to help those numbers along, but even knowing what we know now of their skill and success, aboard that airplane as it approached the water, giving up on life was a dangerous impulse. If you don't brace, it doesn't take much of a crash to disable or kill you.

At the opposite extreme were passengers who were ferociously determined to survive. A man in the back shouted, "Exit row people, get ready!" They did. They were. They were six men and six women with four over-wing exit hatches to open at rows 10 and 11. They reviewed the placards and prepared for the race. One of them, a man on the right side, disregarded the need to brace, not because he was resigned to death, but because he had put his hand on the hatch handle in preparation and had to keep watching outside.

Back at 19E, the woman with the baby boy on her lap was beside herself with the order to brace. The man next to her—the athletic, self-confident one—asked if he could brace her son for her. He seemed to her to be there to do it, so she passed her boy to him. He put the baby to his chest, raised one knee against the seatback in front of him, and leaned back hard into his own seat to keep from crushing her child on impact. He did it at risk to himself and his own five children. What more can be said of anyone? A few rows back and on the far side, the baby's father cinched his four-year-old daughter tightly into her seat belt, then leaned over and covered her with his body. The Spanish-speaking woman was doing the same with her daughter.

It is a testament to the qualities of an average crowd that spontaneously and without formal direction the passengers performed so extraordinarily well. This is what officials worldwide have a hard time admitting. Were the passengers like Spam in a can, and just along for the ride? A fifty-two-year-old man in 13B took it upon himself to keep the mid-cabin informed during the final descent to the water. Because he was a boater, he knew the Hudson well. He announced that they would be landing near the *Intrepid*, and that this was good because the Midtown ferry traffic would be nearby. As the airplane approached the water, he yelled out estimates of its proximity to touchdown. He was uncertain at first. He could not judge the variables of Sullenberger's flare, the gentle climb, the critical trade-offs being made between sink and airspeed. He knew nothing of flight envelope protections. But in the end, when the computers were hanging the airplane on the edge of a stall, he got the timing right. This was helpful to the strangers around him. Just before the crash, he yelled, "And impact will be . . . now!"

•

For flight attendants Donna Dent and Sheila Dail, sitting side by side in the front and facing aft, the touchdown felt like a very hard landing. For Doreen Welsh, sitting in back and facing forward, it felt more severe. The difference was real. In the front, the cabin remained entirely intact. In the back, oxygen masks dropped from the ceiling, the floor buckled upward, components broke in lavatories and the galley, and the jagged beam penetrated the cabin from the cargo compartment below, close to Welsh's feet. Beneath the bedlam in the back, some passengers heard the sound of tearing metal as the belly ripped open and the subfloor structures succumbed.

The fuselage was engulfed in such heavy spray that throughout the cabin people in the window seats thought that the airplane had gone entirely underwater. The impression was fleeting. The airplane slowed fast, and in less than 700 feet came to rest floating calmly on the surface. For an instant nobody moved; then pandemonium broke out as people surged into the aisles.

For the purposes of government certification, manufacturers have to demonstrate that their airplanes can be evacuated in ninety seconds or less with half of the exits unusable. The demonstrations must be performed live, by people without special training. Typically they are volunteers from among the manufacturer's office staff. At least 40 percent are female, 35 percent are older than fifty, and 15 percent are both female and older than fifty. No children are used, but three life-size dolls are carried to represent babies. The test passengers are assigned random seats in the airplane, which is parked in a darkened hangar and abutted by platforms that can accommodate the escapees safely in lieu of inflated slides. The aisle is level, but littered with the usual contents of overhead bins. The passengers receive a standard briefing from trained flight attendants. They fasten their seat belts and sit there for a while with the shades lowered, until the emergency lighting switches on and they are told to get out. The evacuations that ensue are orderly and smooth, with the test passengers moving politely up the aisles, without shoving or clustering. As a result of their good behavior, they always succeed within the official time limit, and the airplanes are approved.

But the demonstrations are obviously unrealistic. The world's preeminent specialist in passenger behavior, a British researcher named Helen Muir, ran a series of test evacuations that made the point clearly. She started with evacuations much like those used for aircraft certification: as expected, the volunteers behaved well and the evacuations proceeded smoothly. She then made one small change to the protocol: offering a minor cash reward to the first 50 percent of test passengers to escape. The effect was to gum up the evacuations, and in the most dramatic way. Muir filmed the scenes for science. Inside the cabins, social norms completely broke down. The same people who previously had been so civilized now stampeded up the aisles, shoved others out of the way, went climbing over the seats, jammed into writhing clusters around the bulkheads and obstacles, ejected flight attendants from the doors, and generally pushed so hard that Muir's staff had to save people from being trampled and crushed. And that was just for cash. In the real world of airline accidents, when panic drives people toward the exits and the competition is about survival, the breakdown can be even more severe—and evacuation times correspondingly longer. Small riots have broken out in certain cases, with passengers fighting each other or violently resisting the crew, and the dead later found piled around the bulkheads and doors.

For their part, the occupants of US Airways 1549 were powerfully motivated to compete. They did not face the ultimate terror of flames, but they were jammed into a sinking airplane that many feared would at any moment explode. Some people panicked and did foolish things, and one man is reported to have shoved a woman down in the aisle in order to pass by. Nonetheless, it is a testament to that average crowd that another man helped her up, and that reason generally prevailed. The flight attendants also performed well, particularly in the front. Dent went to the left door, and Dail to the right. After peering through the prismlike portals and realizing to their surprise that they were in the water, they saw that the nose was sufficiently high, and they swung the doors open fast. Because the nose was high, the bottom sills stood about two feet above the river. Dail's slide raft on the right side inflated automatically. Dent had to struggle with the slide on the left. While she did, the first passenger appeared by her side and, uncertain what to do, took off his shoes and jumped into the frigid water. The second passenger waited for Dent, and shot a look at a man behind him who was yelling, "Go! Go! Go!" The raft inflated, and he jumped in. Others followed in quick succession, one by one, and without shoving. The man in the water swam toward Manhattan for about a minute, then swam back to the raft and crawled in. This was a wise move. Immersed in the cold water, he would have lived fifteen minutes at most.

At the right door, the evacuation proceeded apace.

Back in the mid-cabin, passengers had moved even faster, popping the over-wing exits well before the front doors opened, and beginning to pile out onto both wings. For the moment the left wing was dry. The first two passengers onto it jumped into the water anyway, because they assumed the airplane was rapidly sinking. They swam a few strokes to the forward slide raft, which had just deployed. The third passenger jumped into the water behind the wing and drifted as far back as the tail, before returning to crawl into the small off-wing slide, which had twisted at first, but by then had been sorted out. Soon enough, people realized it was probably better just to stand on the wing, even as it subsided below the surface, than to brave a swim in the winter waters. Ultimately, thirty-six passengers were rescued from that wing, and eight from the off-wing slide.

The right wing was below the surface from the start, and by the end the water there was waist deep. The first passenger out was the man who held the exit hatch handle rather than brace. Soon afterward came a woman who slipped and fell into the river. The man went in after her and brought her back to the wing.

Inside the cabin and toward the back, the situation was worse. Water flooded in so fast that it was up to people's calves before they could even get their seat belts released. The woman who had counted during the departure had closed her eyes for the crash, but now she bounded into action and went climbing forward over the seats. When she got to the exit rows, the wings were already so crowded that she kept right on going, all the way to the front, and made it into Dail's raft. This appears to have been the seat-climbing record for the event.

The woman with the lap baby found herself in the aisle with no memory of having gotten there. The man who had shielded her son handed the baby to her. The baby had not even whimpered during the crash, but he was crying now. She took him in her arms and, because the aisle was blocked with people, began to climb the seat backs. This was difficult to do while holding her child, and she had to stop after five rows. While she hesitated there she witnessed a woman trying to extract a suitcase from an overhead bin, and another passenger ordering her forcefully to leave it behind. The aisle seemed to her like a stampede. The Californian businessman came along and asked her why she was not taking her baby to the exit. She answered that she couldn't break into the crowd. He placed her and her child in a protective hug and propelled them up the aisle. She closed her eyes until he got her to the exit row, and then she waited until her husband and young daughter arrived. Together the family waded out onto the right wing, where a chain of other passengers helped to secure both children in the off-wing slide. Ultimately, twenty-two passengers were rescued from the right wing, and twenty-one from its off-wing slide.

The scene in the aft galley was the worst. About eight passengers had clustered there, along with Welsh. The river was pouring in heavily through the broken aft bulkhead, and someone—it is not clear who—had compound the flooding by mistakenly cracking the submerged left door. Welsh declared to the passengers that they had to turn around and go forward fast—and that they had two minutes to survive. By then water was rising to chest-high. She started forward herself, and at that time (or possibly before) cut her leg badly on the beam that had pierced the floor. Bloodied and in shock, she made it to the front right raft.

The Spanish-speaking woman and her young daughter had also made it to that raft. Up in the front, Donna Dent noticed that the elderly woman who had boarded the airplane with a walker was trying to climb the increasingly steep aisle and, even with her daughter's assistance, was obviously having trouble. Dent went down the aisle into the water, past the over-wing exits, and helped them forward to the right door, before returning to her station on the left. Soon afterward the last of the passengers boarded the rafts.

Sullenberger and Skiles had been assisting throughout, moving through the cabin, reaching under the seats to hand out life vests, helping to direct the passengers; and supervising the evacuation. In the end, when the last passengers appeared to have escaped to the outside, Sullenberger went back through the cabin one last time, making sure it was empty. There were limits to how far he could go. The aft galley by then was completely underwater, and the Hudson had risen to the tops of the overwing exits. Sullenberger came forward again, ordered Dail into the right slide raft, and knelt to use the quick-release mechanism and free the raft from the door. About five minutes had passed since the touchdown in the river, and already the first rescue ferry was standing by on the right side. On the left side, Dent climbed into the remaining raft, followed by the pilots. Sullenberger was the last off the airplane. He sat next to a passenger, who thanked him. Sullenberger said, "You're welcome." He didn't say much else. His demeanor was calm, as it had been throughout. Skiles was heard to explain that they had hit birds, and could not have avoided them. Within a minute a second ferry arrived there, on the left side, and someone on its deck passed a knife to the passengers, who passed it to Skiles, who used it to cut the raft's tether to the airplane. Skiles dropped the knife into the river to preclude puncturing the raft. People began to climb onto the ferry, some with difficulty because they had grown weak from the cold. A police helicopter hovered overhead, and dropped two rescue swimmers, but complicated matters with its downdraft. Other boats arrived, and people began to swim and clamber to safety from both wings. Sullenberger was the last to leave his raft. Once aboard the ferry he gave his uniform jacket to a man who had been in the water. During the short ride to the shore, he remained mostly silent. He has never said whether he looked back at the ruined airplane floating in the river behind, or whether he felt any gratitude for its performance. At that time it was not known how many people had survived. He had tried to organize headcounts in the rafts and on the wings, but none had succeeded. During the ride to the shore, and for a while afterward, he had to consider the possibility that some of the passengers had drowned. It was a while before word came that everyone was safe.

By official count, thirty-three passengers had put on life vests from under the seats, and twenty-nine people at one point or another had swum. In the end, sixty-four people were rescued from the forward slide rafts, twenty-nine from the off-wing slides, and fifty-eight from the wings. This tally leaves the position of four passengers unaccounted for, who nonetheless survived. The last passenger was rescued from the airplane at 3:55 p.m., twenty-five minutes after the landing. One-third of the evacuees were transported to New Jersey, two-thirds to New York. Some went to hospitals for checkups, but many did not. Some felt that the accident had changed their lives. Some simply returned to LaGuardia to catch later flights. Soon enough the union and the NTSB descended on Sullenberger and Skiles. Meanwhile word of the accident had spread electronically, and thousands of residents and office workers had gone to the shoreline parks or stood at riverfront windows to view New York's latest strange sight. Many believed that they were witnessing a miracle of some kind. This may or may not have been right, depending on how the word is defined. But what they undeniably did see was an Airbus in the Hudson, drifting nose high like a beast in the water, and refusing to die.

He just wanted a decent book to read ...

Not too much to ask, is it? It was in 1935 when Allen Lane, Managing Director of Bodley Head Publishers, stood on a platform at Exeter railway station looking for something good to read on his journey back to London. His choice was limited to popular magazines and poor-quality paperbacks – the same choice faced every day by the vast majority of readers, few of whom could afford hardbacks. Lane's disappointment and subsequent anger at the range of books generally available led him to found a company – and change the world.

'We believed in the existence in this country of a vast reading public for intelligent books at a low price, and staked everything on it'
Sir Allen Lane, 1902–1970, founder of Penguin Books

The quality paperback had arrived – and not just in bookshops. Lane was adamant that his Penguins should appear in chain stores and tobacconists, and should cost no more than a packet of cigarettes.

Reading habits (and cigarette prices) have changed since 1935, but Penguin still believes in publishing the best books for everybody to enjoy. We still believe that good design costs no more than bad design, and we still believe that quality books published passionately and responsibly make the world a better place.

So wherever you see the little bird – whether it's on a piece of prize-winning literary fiction or a celebrity autobiography, political tour de force or historical masterpiece, a serial-killer thriller, reference book, world classic or a piece of pure escapism – you can bet that it represents the very best that the genre has to offer.

Whatever you like to read – trust Penguin.